Learning through Geography

Learning through Geography

An Introduction to Activity Planning

FRANCES SLATER
University of London Institute of Education

 Heinemann Educational Books

Heinemann Educational, a division of
Heinemann Educational Books Ltd,
Halley Court, Jordan Hill, Oxford OX2 8EJ

OXFORD LONDON EDINBURGH
MELBOURNE SYDNEY AUCKLAND
SINGAPORE MADRID IBADAN
NAIROBI GABORONE HARARE
KINGSTON PORTSMOUTH (NH)

British Library Cataloguing in Publication Data

Slater, Frances
  Learning through geography.
  1. Geography – Study and teaching
  I. Title
  910'.7'1   G73

  ISBN 0-435-35715-8

Printed in Great Britain by
Butler & Tanner Ltd, Frome and London

'I still had my gift, a pair of earrings, tiny, world-shaped silver pendants with the lines of latitude and longitude marked like spiderwebs, enmeshing the world: no land, no sea; a smooth shining silver blank surface of the world.' Janet Frame (1981) *Living in the Maniototo*, The Women's Press/Hutchinson.

Contents

Acknowledgements

This book is one of the outcomes of my work within the geography department of the University of London Institute of Education.

I wish to acknowledge that the opportunity and initial impetus to write the book came from Professor N. J. Graves. It has been a pleasant and stimulating experience to work in his department over the last seven years and it has done much to expand my concept of geographical education.

I also enjoyed and benefited from being able to discuss the framework for my book with Professor Nicholas Helburn during his year as a visitor in the department.

I express thanks to John Wolforth and John Huckle for their comments on Chapters 1 and 4 respectively; to Terry Hearn for useful advice; to a non-geographer, Eleanor Walsh, for introducing me to an expanded notion of teaching, outlined briefly in Chapter 5; and to John Fien for his willingness to discuss enthusiastically matters pertinent to a range of issues in geographical education. I should not neglect to mention either, the benefit one derives from students who, through comment and question, criticism and appreciation, provide that essential tension needed to take one's thinking and efforts further.

Like all of us, my ideas have been influenced and shaped continuously by people too numerous to name but who are to be found around the globe and more particularly clustered in New Zealand and Australia, the United States, Canada and Western Europe and to whom in the course of a career, one is grateful. They are often met at conferences and appear in 'the references'.

The emphases and, perhaps, quirks in this book are, however, of my own making. These remain after Joan Rose most skilfully and patiently translated handwriting to typewriter with goodwill and good humour.

Frances Slater
London, 1981

The author and publishers would like to thank all those who have given permission to reproduce extracts, tables and illustrations as indicated in the sources.

Preface

This book is about lesson planning for learning through geography. It is addressed to those who are about to take up the task for the first time and to those who may read it in the spirit of reflecting upon their established practice. In both cases, I am aware that there will be some acceptance and some rejection of the ideas and arguments I put forward.

Teaching is a highly personal and somewhat idiosyncratic activity. Our individual personalities have a strong influence on our style of planning and presentation. Yet we all enter our classrooms with a plan in mind. The plan may be detailed or sketchy. Sometimes, we have a great deal of success with plans we consider sketchy and less with those we have taken a long time over. Sometimes the reverse is true. The essential point is that we do have plans, schemes or structures in our minds which guide and inform our practice to give us satisfactory or unsatisfactory outcomes, however we define satisfactory or satisfaction.

Our plans and guidelines, whether intuitively or consciously operated on, sometimes benefit from another's point of view. It can be stimulating as well as painful, in the light of another's statement, to reflect upon, even to rethink, one's own gradually developing or well-developed set of ideas about the way we can best go about a task.

My approach to lesson planning (or my preferred term, activity planning) is based on four elements. These four elements which intermingle throughout the book are: my own common sense, experience, and preference; Nicholas Helburn's (1968) activity analysis models of the process of teaching geography; some of the suggestions for planning units of work put forward by the Secondary Geographical Education Project of the Australian Geography Teachers Association of the state of Victoria; and general and particular concepts in the field of curriculum theory and development.

I think the combination of elements has led me to set out an approach to planning lessons which both

places students as well as teachers in the role of data gatherers, processors and interpreters of their own and other people's experience and yet does not negate or play down the teacher's role as overall organiser and sensitive monitor of a learning process in which he or she is enthusiastically engaged.

To that extent, readers will appreciate that the book is not about curriculum planning in a broad sense but about preparing lessons by selecting resources and choosing teaching strategies.

The curriculum planning process in geography has been specified (Biddle, 1976; Graves, 1979) at three levels which progress from very general to very particular contexts. If we always saw things logically and if, indeed, things did develop in a logical fashion, then prior to lesson planning we would have been engaged in decisions about the aims and make-up of the geography curriculum in general. These decisions would guide and inform practice throughout a country, a school system or a school. These general decisions would then have to be translated into programmes of work suitable and appropriate to early, middle and later years of schooling.

It is at this programme planning level that most, though by no means all, geography teachers come into contact with curriculum planning. If fortunate, they will be given a curriculum which makes general suggestions on what topics, themes, regions or problems are to be studied at first year, second year and so on. With reference to such a curriculum document, resources have to be selected, key questions set up and lesson activities organised. It is precisely the patterning of learning activities centred on topics, regions, problems or themes with which I am concerned in this book.

The pattern which I suggest for planning learning activities is composed of identifying questions to reach generalisations through data processing and interpretation. The next five chapters are an elaboration of what I mean by identifying questions, reaching generalisations, data processing and data interpretation in the context of planning the sequence of a learning activity to promote learning through geography as a worthwhile component in a person's general education. Furthermore, I undertake the task in the belief that the quality of our planning in the past and in the future is one of the most significant variables determining a continuing role for geography in education.

1 Identifying Questions to Plan Learning Activities

The role of questions in planning

In this chapter, two main ideas are woven together. The first is the practical utility of identifying questions and sub-sets of questions as the initial task in organising and planning learning activities. *Guiding* questions must be sorted out. Once this is done the direction and selection of teaching-learning strategies and resources for the activities can be decided. Question formulation is emphasised throughout the chapter and several sample teaching-learning activities are worked through. These activities contain facts, ideas and issues currently acceptable as worthwhile geography. It should be noted that identifying questions does not of itself create intrinsic or extrinsic motivation. Arranging the episodes and sequencing activities to engage the interest of learners has to be considered after the broad key decisions of what questions, what data, and what generalisations have been taken.

Science and humanism

The second idea is the notion that 'geography as science' and 'geography as personal response' to the environment both have a part to play in developing student understanding through activities based on geography. Both logical positivism and humanism have a contribution to make to geography. 'Geography as science' is characterised by its abstraction and modelling of reality, and it has made tremendous contributions to the development and rigour of geography in schools through its power to engage and strengthen reasoning skills. 'Geography as personal environmental response' directs attention to our experiences and interpretations of everyday life whether structured cognitively or emotionally, but more especially, emotionally.

What is subjective and personally meaningful in our knowledge of places and spaces, and what is apparently objective, analytical, spatial enquiry are not juxtaposed in this and subsequent chapters but are presented as complementing one another. The complementarity is illustrated by examples of learning activities based on both personal preferences and feelings and also on objective knowledge.

Geography in a psycho-analytic age

Geography as science needs less defence or explanation than geography as personal response. The former is for the most part well-established in forward-looking educational systems and schools and should continue to play an important part in a geographical education. Geography as personal response perhaps needs a little more justification if it is to be given a chance to provide a necessary complement to geography as science.

Paul Claval (1978) seems to have a telling argument which strengthens the case for humanistic geography in schools.

Claval states that in French secondary education, the aim of geography for the last two centuries has been to teach children to locate themselves within their cultural framework, i.e. the framework of greco-roman civilisation. Such a conception of education as initiation into a culture has stood for two centuries, with geography and history being accorded a place in the curriculum as cultural disciplines.

Today, however, there is a questioning of the ideological underpinnings of western societies which Claval feels has been opened up by concepts in psychoanalysis. Classical people located themselves in a broad society, by knowing its history and centring themselves in the larger world. Psycho-analytic humans do not look so far into time or space. They find keys to their evolution close by, in the society of the nuclear family. The problem is not now for western people to integrate themselves into a continuity of time and space but for individuals to build universes to their own measurements, as detached as possible from collective values and objectives.

It is at this point in Claval's analysis that I would assert that geography as personal response has a part to play in the geographical education of psycho-analytic humans. Humanistic geography offers possibilities for individual experiencing and meaning-making, for elucidating personal introspection and emotional bonds with places.

Whether the learning activity planned is akin to geography as science or geography as personal response or a combination of both, a question identification approach is a feasible and practical opening strategy in lesson planning in geography.

Question identification

A philosophical argument developed by R. G. Collingwood (1939) points to the necessity to preserve a direct link between questions and their answers. (For an earlier rehearsal of this line of thinking see Slater, 1975.) He holds that this is necessary if an intellectually satisfying understanding of an issue, topic or problem is to be achieved and maintained. This leads us to the assumption that in order to arrange and plan lessons or learning activities, the identification of a number of specific questions is necessary to provide guidelines for the ordering and selection of concepts and content within an activity or series of activities. The case for a question identification approach to activity planning is modelled on one which Collingwood constructed in order to expose what, to him, was the false position of propositional logic. Collingwood maintains that the questioning activity in knowledge is of fundamental importance:

> . . . A body of knowledge consists not only of 'propositions', 'statements', 'judgements' or whatever term logicians use in order to designate assertive acts of thought (or what in those acts is asserted for knowledge means *both the activity of knowing* and *what is known*), but of these together with the questions they are meant to answer. . . . (Author's emphasis)[1]

Collingwood illustrates his argument as follows:

> . . . If my car will not go, I may spend an hour searching for the cause of its failure. If, during this hour, I take out number one plug, lay it on the engine, turn the starting handle, and watch for a spark, my observation 'number one is all right' is an answer not to the question, 'why won't my car go?' but to the question, 'is it because number one plug is not sparking that my car won't go?' Any one of the various experiments I make during the hour will be the finding of an answer to some such detailed and particularised question. The question, 'why won't my car go?' is only a kind of summary of all these taken together.[2]

Key questions and sub-sets of questions need to be identified in a Collingwood type of searching activity to get lesson planning under way. When questions have been identified a teacher can fashion an activity which step-by-step leads students towards an understanding of a problem or general idea. Collingwood argues that there can be no real understanding unless the *answers* can be *directly* correlated to the *questions* and, after all, to achieve understanding is a generally accepted overall goal, if not a definition of, any learning activity (whether the activity is one of several blocks building towards a generalisation, or the ultimate block in that process). The imperative for the maintenance of a close link between questions and answers suggests that knowledge and understanding are incomplete when the generalised answers alone and in bulk are presented to learners, whether these derive from research articles or textbooks, dictated notes or flowing expositions. 'What is she/he getting at?' is a familiar response to an activity in which the teacher has not made *explicit* the link between the question being explored and the answer or conclusion being presented. And the two must be clearly identified in the very first stages of planning. Failure to make the question—generalisation link explicit to students usually means that students miss the point. If the point has been missed, how can new learning be related to old learning?

Gagné's (1970) hierarchy of intellectual skills has something to contribute to this argument. Gagné reminds us that success in learning depends on certain prerequisites being met. In a much broader sense, our planning must specify all necessary prerequisites, questions and related answers if a framework for planning learning is to be set up.

Product not process

Consider the regional paradigm and how it often came to be applied in schools. Regional studies were originally developed in order to answer a question which went something like this: 'What are the inter-relationships among phenomena that produce this particular set of features?'

If this question is selected for investigation, then the regional paradigm provides a means of demonstrating the inter-relatedness of phenomena from place to place and an understanding of the factors which contribute to an areal differentiation of the earth's surface. The question of whether geographers had the tools to seek out and identify all the relevant inter-relationships is left to one side, though the rise of the regional analysis school suggests that the technological and conceptual means are now available. In regional studies, very often the whole emphasis came to be placed on the distinctive set of features and not so often on the inter-relationships. So, in translating the use of the regional paradigm into schools and university teaching, what sometimes happened was that geography in the classroom and the lecture hall in its extreme form became characterised by the description of areas, by the compilation of inventories of the content of areas with, frequently, little emphasis on explanation.

The process of investigation, of asking questions and developing lines of reasoning, became totally separated from the conclusions which alone were presented. The learning of hundreds of facts placed potentially meaningful material in danger of being learnt by rote. The separation of *guiding* question and answer deprived students of a source for structuring their learning material and cognitive activities.

Psychologists like Ausubel, Bruner and Gagné suggest in their different ways that a student needs hooks on which to hang new knowledge, that new learning needs to be related to the old. Indeed, not making clear the link between question and answer in our planning hinders the creation and development of learning hooks. Examinations in regional geography tested essay writing skills and required the ability to organise material but not necessarily critically to analyse or evaluate it. Indeed, given the world-wide coverage which many syllabuses demanded, teachers and students were forced to compile checklists of the relief, climate, agriculture, industry and settlement characteristics of places. The view that the uniqueness of each region required a knowledge of all regions all taught in a similar style produced a state of underdevelopment in critical and logical thinking skills in the geography classroom. Such was one of the consequences of the separation of question, process of investigation and answer and the adoption of a set pattern of teaching.

Exactly the same coagulating, deconnecting process could happen in modern geography courses based upon thematic or conceptual structures. Consider what *really* happens in a lesson on settlement location when activity is limited to students taking notes on:

1. Position and relation to routeways.
2. Influence of resources.
3. Historical reasons.
4. Site and situation characteristics.
5. Chance elements or other factors.

The danger of both teacher and students failing to see and develop the connection between a question and an answer is only too real.

Questions initiate. What questions?

The recent reorientations of geography and geography in education with the identification of fundamental organising concepts, especially those related to spatial organisation or locational analysis viewpoints, have been paralleled by fresh approaches to classroom teaching. These developments could be in vain if the leap towards conclusions is made at the expense of tying such answers to the starting point—the question—which provided the initial impetus and guide. Teaching and learning are question-asking activities and planning these activities seems to proceed most readily after initial questions have been identified.

Very often the first in a series of questions on a topic will contain one or more of geography's major concepts, e.g. Why are settlements *located* where they are? Is there a *pattern* to the distribution of cities in England, Japan and Canada? What *connections* (or inter-actions or links) are there between settlements? Has the *distribution* of settlements changed through time? Where should a new settlement be *located*? Are there different *patterns* of land use in a town? Why?

What processes are at work? Are some places more *accessible* than others? Why? Why do settlement *patterns* and *densities* differ throughout a country or the world?

Within any of these big questions (big, because they contain concepts fundamental to geographical studies), there are sets of other questions which can help to structure an activity or unit of work towards understanding and generalisations, as we shall see later in the chapter. Such a sub-set of questions for 'What is a settlement?' might include 'What characteristics of towns and cities do people prefer? What do people like best about their towns and cities? Is it helpful to classify settlements by size, function or desirability?' These questions pose problems or intellectual tasks of *identification, definition, description, classification*, and *analysis* which help to *explain* and answer the big question, 'What is a settlement?'

Concepts and skills

The intellectual skills brought into action to identify and define, describe and explain in order to tackle the 'big' question involves students in processes of thinking, rather than simply and solely in the digestion of the products of other people's thinking.

Practice in the skills of observing, defining, classifying, analysing, inferring and so on helps train students to transfer such processes and procedures of working to new problems and new questions. Students are thus provided with firstly, a template of the concepts used in geography (e.g. location) and secondly, the processes involved in thinking (e.g. observe, define) which go beyond, and, indeed, link any number of specific questions. Essentially, most questions will be concerned with either what and where things are or how and why they are where they are.

Questions as signposts

Early models of curriculum development suggested that the first stage in activity planning should be the determination of objectives. Taba (1962), for example, defined objectives as 'paths to follow'. However, it now seems clear that identifying questions gives us signposts to follow in activity planning and helps us to be more responsive to situations as they arise in

the classroom. Signposts give general directions while we recognise that we may discover other paths to learning in the process of learning. In teaching, we need to recall continually the questions on the signposts and to remind ourselves why we are going where we are going, but changing the route, even the destination, if student attention or level of understanding indicate this is necessary.

SGEP

This advocation of an 'identifying questions' approach to activity planning parallels recent experiences and practices in the Secondary Geography Education Project (SGEP) (1977) of the Geography Teachers Association of Victoria, Australia. The British Schools Council Geography 16–19 Project is also giving a significant place to key questions in planning. Doubtless, other undocumented curriculum groups and individuals have adopted similar methods. The experience of the Australian teachers and the very successful results of their curriculum planning provides welcome evidence of the viability and practical nature of the 'identifying questions' approach to lesson planning. Equally important for practical planning, the strategies they used are valid, replicable working methods for any teachers planning activities or engaged in curriculum renewal. For these reasons it is worthwhile to review briefly the origins and development of the Victorian Secondary Geographical Education Project.

SGEP guidelines

Externally assessed examinations ceased in Victoria in 1971 below form 6. Teachers were given complete freedom to design their own courses for years 7–11 (that is forms 1–5 in England). Some years later evidence indicated that teachers wanted guidance in unit and curriculum planning but not a return to examination prescriptions. SGEP was designed to meet their needs in four stages through:

1. Developing and disseminating course construction rationale.
2. Arranging courses.
3. Developing and disseminating exemplar units of work and resources.
4. Developing and disseminating teaching/learning strategies and styles.

The initiative to develop materials in the latter three stages was placed on regional and local groups of teachers. From the beginning, teachers at SGEP meetings and conferences supported the idea of what came to be called 'enquiry' approaches to course planning. An enquiry approach in the SGEP context means a course consisting of a sequence of 'units' (learning activities) framed by geographic questions where each unit is broken down into a series of key *questions* which place emphasis on what students will actually *do* in the unit. The difference between the enquiry or question approach to course planning and more traditional thematic, topic or skills based approaches is illustrated in the following example from the SGEP publication. Consider the following:

1. 'poverty'
2. 'Are there "pockets" of poverty? Why?'

In the first case, it is not clear from the title exactly what students will be expected to learn about poverty. In the second case, there is some idea of what students will be expected to do and an indication of where the study is leading. Sample sub-questions which apply to this enquiry (and most others in geography) might be:

Where is it? Where does it occur? What is there? Why is it there? Why not elsewhere? What could be there? Could it be elsewhere? How much is there at that location? Why? How far does it extend already? Why? Is there regularity in its distribution? Why? Where is it in relation to others of the same kind? What kind of distribution does it make? Is it found throughout the world? Is it universal? Where are its limits? Why? What else is there too? Do these things usually occur together in the same area? Why? Is it linked to other things? Has it always been there? How has it changed spatially (through time)? What factors have influenced its spread? Why? What is the area likely to become? Why? How should the area be used?[3]

Bernard Cox more succinctly lists as key questions for geographical investigation:

1. Where are things located?
2. Why are they there?
3. What are the consequences of their location?
4. What alternative locations may be considered in decision making?

These are used as guiding questions in Queensland geography syllabus outlines.

Generating questions

Throughout the development of SGEP courses, *brainstorming* for questions and ideas was a key activity. Brainstorming was a significant element in the American HSGP development strategy also. The aim of brainstorming is to create as many ideas as possible, paying at first little attention to their worth. Criticisms and evaluations are suspended until divergent lines of thought are exhausted. Combining and recombining the suggested ideas produces a list of possible unit titles and learning activities. The SGEP approach recognises the value of brainstorming as a procedure to be used with small groups and recommends that school geography departments use brainstorming as a starting point in their course planning. Once general areas of work have been decided upon, each area should be structured into an 'enquiry', as a set of questions. This serves to emphasise questions acting as signposts to the nature and direction of the learning activities.

Classroom activity

Where is settlement likely to develop?

The strength and utility of the question identification approach to activity planning now needs to be demonstrated. For this purpose, the settlement siting activity from *Geography of Cities*, Unit 1 of the High School Geography Project (1969, 1979) has been chosen, since it is a good example of the question identification approach centring on the main question, 'Where is settlement likely to develop?'

It should also be noted that the exercise below has been amongst the most widely disseminated and successfully used and remodelled of the many American High School Geography Project materials. Its teachability is well proven and it has been taken up by social studies teachers in the United States and Canada. (*Experiences in Inquiry*, 1974.) It appears redesigned in textbooks outside North America (see, for example, Knight, Buckland and McPherson (1973)). Hall (1976) should also be consulted for his analysis of and comment on this exercise.

As a springboard for this exercise, it helps to have students think about the location of their own settlement, through such questions as 'Is it in the middle of a mountainous region? In a desert? Is it near a river?' The answer to the first two questions is likely to be 'no' but to the third 'yes'. Of course, it is possible for settlements to develop in mountains and deserts. Communities have been built in Greenland and Alaska and temporary settlements exist today in the North Sea. Most often, however, settlements have grown up in river valleys, on plains or undulating country and along coast lines. As a quick check ask students to see how frequently such locations occur for ten of the world's largest cities. A question to follow the initiating activity, which is judged to be motivating, could be 'If you were one of a group of new settlers in a country what kind of things would you look for before deciding where to set up a camp?' The teacher should decide time and setting. For example, a student could be a Scotsman settling in New Zealand in 1840 AD or one of a group of Saxons coming to England in 500 AD. Typical student responses include the ideas that settlers would have to be able to farm and grow food, obtain water, fuel and building material as first priorities (see also Chapter 5); that settlers would be attracted to sites which they hoped would have fertile soil (How would they know whether the soil was fertile?). Similarly, if they thought they could take up some occupation such as timber milling, they might settle near places where supplies could be brought in with little difficulty, either by land or water and where connections for sending out the timber to other areas were reasonable. So a sheltered harbour becomes a top priority guiding choice. Some students are quick to observe that people often modify sites by draining swamps, filling in tidal lands, dredging rivers, levelling hills and building flood protection walls. Tracks, roads, highways, canals and railroads can be developed in the course of time to improve connections with other places. How might some priorities change if the students were told they were a Norman baron responsible for defending an area or, similarly, a settler in hostile Indian territory?

Choosing a site

After such introductory discussion, students should examine Figure 1.1 in a North American context and answer the following question for each sketch: 'Where is settlement most likely to develop (or locate) in the year indicated?' Responses to the question relating to the sketch for 1800 have been found to take the following lines: Site B is close to the ocean and ocean-going ships. However, it may be vulnerable to wind and storm waves and it is very close to low-lying land. Although on a river, site D is relatively inaccessible and site C is also isolated, but perhaps easily defensible. There would be little room for the expansion of a settlement if settlers happened to think in terms of the future. Site A might be chosen

for its location at the confluence of two rivers or rejected as a site liable to flooding. These and other suggestions, hunches or hypotheses provide feasible reasons for deciding among sites A, B, C and D as a place to settle.

Any additional information which a teacher invents about the sites will eventually tip the balance in favour of one. Unless a firm conclusion is wanted such a strategy need not, and is probably best not, built into the exercise. It was an HSGP general educational intention that students should experience and come to tolerate some ambiguity. It is not the site chosen but the quality of the reasons for a choice which is most important. An ultimate decision, however, could be taken by a show of hands (indicating preferences) or a throw of dice — collective decision making and chance, after all, explain many settlement locations. Further justification for leaving the lesson open-ended, depending on the age and background of the

Figure 1.1　Selecting sites for settlements

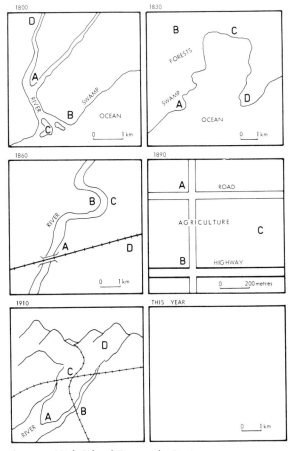

SOURCE: High School Geography Project.

students, is to have them consider the extent to which first settlers might have short- or long-term priorities and might have a comprehensive or perfect knowledge of the site, its surroundings and future opportunities in the area, e.g. a discovery of iron-ore could favour or discount the initial choice. This kind of initial activity could lead students into a study of the history of their own settlement or some other settlement for which detailed documentation on origin and growth exists. In that case, it is interesting to compare a list of the points which students wanted clarified in relation to any of the sketches and the kind of historical detail and fact which is actually available.

Expanding concepts

In the second sketch, labelled with the year 1830, site A may offer a good harbour at a time when ships were a major means of transport and communication. Site B is within the forest, presumably exploitable but well away from the coast. At this point, the concept not only of location but also of relative location can be made explicit. Thus, in the Brunerian sense (Bruner, 1960), the concept of location is given additional depth, breadth and meaning. Site B would require road links with the coast if timber were to be exported by sea. Such construction costs may be worthwhile if sites C and D are discounted. Site D appears to have an exposed location, though it may be defensible. Site C lies at the head of a bay and the mouth of a river.

Other attributes significant to location can be seen with reference to the other sketches. In 1860 the coming of the railroad added a new variable to the selection of a settlement site. This, of course, would modify many previous site constraints. By 1890 rural industries had been established and service centres in agricultural areas had developed. In the 1910 sketch, which facility dominates—rail, river or coastal transport? A tempting final question, future oriented, would be 'What reasons might, should or ought to inform our choices of a settlement site if we were to discover a new continent tomorrow?'

Using questions as a focus

By using questions as the key focusing mechanism on relatively simple settlement puzzles, the link between the question and reasons or answers has been directly preserved. In more complicated exercises, a more detailed set of sub-questions would be necessary with a balance between closed questions (the answers to which are generally fixed) and open questions (the answers to which may be speculative and promote an ability to go beyond the information given).

Types of questions

Five types of questions are suggested as shown in Figure 1.2. These could be useful as a guide in framing and balancing the proportion of closed and open questions asked within the ebb and flow of the classroom activity. In addition to the key questions, normal prompting type queries may be made by a teacher, such as: 'Why do you think that? Can you give another reason for your choice? Do you think "X" or "Y" could be important? Why do you agree/disagree with the previous answer? Is it possible that . . .? What do you mean by . . .? Let's look at the characteristics of site A again. Have you forgotten the importance of moving things by water in 1800? Why is "Z" more important . . .? In other words . . .? Do you really think that . . .?' What is suggested here is that the questioning approach be consciously built in as an essential guide to the nature and direction of the learning activity. Also needed will be such questions as 'Which sketch map are we looking at? What date is it given? What answer has just been given?' for those students whose attention wanders and for whom enquiry becomes wearisome.

From questions to concepts and generalisations

In using the same question in the settlement location enquiry 'Where is settlement likely to develop?' and in making evaluations of all the sites and considering them in relation to the different time periods, students are being guided towards forming generalisations or 'theories' about the variables influencing the decisions to settle at place A rather than place B. If, as is likely, students go on to link the level of transport technology with other variables, it may be considered that a principle or recurring relationship between two or more variables has been established. Linking ideas like this produces generalisations, and learning activities structured to this end are discussed more fully in the next chapter.

In addition, a notion of settlements as central places for goods and services in agricultural areas often emerges in class discussion. The specialised function of settlements as route centres or break of bulk centres, as tourist, forestry or mining centres is a related general concept. For example, looking back to Figure 1.1, site D in 1910 may have developed as a

Figure 1.2 Encouraging critical thinking through questions

Closed ◄——► Open				Critical Thinking
Demanding recall	*Encouraging classification and ordering*	*Encouraging the use of data to draw conclusions*	*Encouraging awareness of the limitations of the evidence or evaluation of evidence.*	*Encouraging an awareness of the processes of reasoning to be used*
What were the dates labelling the 5 sketches? To what extent had the area been settled? What kinds of transport were available?	Closed example: Can you make a list of those things which seem likely to be most important and not so important in site selection? Less closed: Can you think of a way of sorting out the types of settlements that probably developed?	Do you consider the nature of the coastline may have influenced where the settlers settled? How do you know? Can you think of some examples from the real world to support your answer? What influence may the meeting of a river and railroad have had on settlement? How do you know? Why do you say that?	What do we know for certain about where people settle? What are we not so sure about? What do we know for certain about the sites we selected for settlement? What are we not sure about?	How did we go about deciding which was the best place to settle? What different kinds of evidence or information did we use? Were some pieces of information more useful than others? What effect might more information have on our decision about the best place to settle?

Source: based on Blyth, W. *et al* (1976) *Curriculum Planning in History, Geography and Social Science*, Schools Council/Collins.

tourist centre, depending on the level of economic development and the needs and demands of the population. In parts of Europe, the area undoubtedly would have been developed already by that date. In the case of the growth of settlements in the Shetland Isles, for example, the related idea or concept that change in the predominant function of a settlement can occur over time may be developed. Change, of course, can lead to the decline as well as growth of settlements. To these possible conceptual developments, another may be added. Given the layout of the sketch for 1890, the idea of a settlement hierarchy may occur. In 1890, site B seems the most likely place for the development of a major settlement if site C is considered to be an isolated farmstead with perhaps site A as a minor town. The concept of a hierarchy of settlements can be developed from the single starter resources which Figure 1.1 represents, using suggestions or questions made by members of the class.

Making connections

The original question posed, 'Where is a settlement most likely to develop in the year indicated?' in relation to Figure 1.1 is as specific as Collingwood's suggestions demand. The concept of location is not explicitly used but it is the basic and underlying concept being explored. Students can be expected to provide very specific answers and a range of hypotheses. Responses will enable an understanding to be developed of the concerns which may be uppermost in the minds of settlers in new environments. The variety of suggestions, the range of possibilities for correct answers, yields an opportunity to develop generalisations. These might include the effect of new inputs into the system, such as developments in science and technology and their role in influencing people's decisions on what constitute suitable places for settlement. Generalisations about the size, spacing and hierarchy of settlements often evolve from the initial question and the student is enabled to make connections between hitherto unconnected concepts or ideas. This fulfils a Brunerian dictum that any way of structuring knowledge for learning should have the power to make students capable of realising new connections.

An activity like the settlement siting selection lies at the very core of the scientific approach to geography adopted by those working in the locational analysis or spatial organisation paradigm. The central and related concepts of location, relative location, distance (separation) and accessibility help to explain the possible or actual arrangement and structure of patterns.

Geography as science and as personal response

In the introductory statement, it was noted that as the 1970s passed an increasing number of geographers suggested alternative constructs of how the world might or should be viewed. New questions which had much to offer to the development of a student's sense of his or her world were put forward. Among these more recent questions are those providing students with opportunities to explore how space is structured. This adds to their understanding and complements the scientific approach. A humanistic approach emphasises the individual exploration of places. Students are encouraged to find their own meaning of places, a meaning which may or may not be shared with others.

There is a strong case to be made for distinguishing personal meaning of a geographical kind from its public meaning and devising learning activities in both. What personal meaning a particular place holds, what is liked and/or disliked about it, and what other places are associated with it, should be contrasted when possible with publicly held meanings. These public meanings include those generalisations about space/place which fit into a geographer's picture of the world, for example those formulations of space and place represented in spatial models.

Meaning-making

The questions around which learning activities can be structured may move from those having a good deal in common with geography as science to those which explore geography as personal response. The kind of answers or responses will depend on the meaning which students find for themselves. Geography as personal response type questions functions as an opportunity for meaning-making through oral, written or other modes of expression. The outcomes depend much more on how students structure their own thoughts and response than on how the teacher structures the exercise. Geography as personal environmental response type questions mirrors Eisner's concept of expressive objectives or expressive outcomes (see Graves, 1975, 1980; Eisner, 1979), where students are provided with educational encounters in which it is not stated or known precisely what the student will learn in terms of formal content but which are judged to be constructive and fruitful for developing personal meaning and understanding. What is important is the enquiry process itself and the opportunity to explore personal environmental knowledge and experience. Two simple questions illustrate these points.

Classroom activity

'Why do you live where you do?' 'What do you like best/least about that place?'

The balance of objectivity and subjectivity is clear in these questions. They require an analysis of an outside real world and an inside personal world. The first question, 'Why do you live where you do?' can have a sub-set of questions:

How long have you lived there?
Has your family ever moved?
Where did you live before you came here?

It can be accompanied by questions for parents to answer, e.g.

Did you choose the house because...
it was close to work?
it was close to relatives?
the price was suitable?
it was the right size?
the appearance was nice?
it was in a good locality?
it was close to a school?
other reasons?

The data collected from students and parents may be analysed to calculate a mobility/stability index for the class. This can simply be the percentage of students who have moved in the last five or ten years or, conversely, the percentage who have never moved. Further fieldwork can be undertaken to survey other classes or, perhaps, residents of streets or apartment blocks in the neighbourhood. The specific reasons for housing choice can be ranked from 'most frequently' to 'least frequently' occurring and histograms drawn. Comparative data can be collected from other classes or neighbourhoods.

Expressing feelings and attitudes

The second rather different question, 'What do you like best/least about the place where you live?' is introduced to give students an opportunity to clarify and express attitudes and feelings they have towards a place or part of it. The key question and its sub-set, could be drawn up as follows:

> What do you like best/least about the place where you live?
> Why do you like it or one part of it best/least?
> Describe what things are in the place. How are they arranged?
> Describe your feelings when you come into your best/least liked place?
> What kind of things do you do there?
> Is it what's in this place or what's not in this place that makes you like/dislike it?

How do you feel when you leave the place?
How do you treat it when you're there?
Can you explain why you like/dislike it?
Does the mood you're in affect how you feel about the place?
Does the place affect your mood?[4]

Here the questions are focusing on the emotional bonds between a person and a space or place which is experienced in ordinary, everyday life. The questions are attempts to help students to become aware that they have feelings and attitudes towards places in their environment. At a deeper level, older students could be encouraged to ask about the nature of the relationships people form with parts of their environment. Their environment could be viewed existentially, i.e. not as that space, place or environment which is there already but the space which they have the power to make and remake every day in their thoughts. From such a viewpoint they may be encouraged to interpret the qualities of various environments and the feelings which they have towards them. To give students the opportunity to create and to be aware of the process of meaning-making provides a rationale for educational subjective encounters suggested by questions like those above.

A very geographically appropriate example drawn from an environmental education package, Essences (1970), is set out in Figure 1.3 which details an exercise requiring students to delineate environmental areas of comfort and unease.

Figure 1.3 Mapping areas of ease and stress

the action: _____

> Map the places in your environment in which you feel the most and least comfortable.

more:

> How about a cat?
> How about a bird?
> How about another person?
> How does your comfort conflict with the comfort of others?

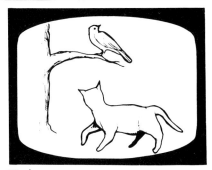

Source: based on American Geological Institute (1970) *Essences 1*, Addison-Wesley.

Private geographies

John Fien (1979, 1980) and R. J. Gilbert (1979) have recently argued that exploring the sense and meaning of a person's felt relations with spaces and places allows students' subjective, and essentially private, geography to be recognised and delineated. Everyone has conscious and unconscious feelings towards, and a sense of how they feel about, what is around them. Awareness and knowledge of one's surroundings are unique to each individual, as we do not all receive the same sensory inputs in the same manner or order. Private, personal geographies based upon the perception and experience of worlds real or imagined, and the resultant environmental feelings and images actually do exist. At the very least, private geographies should be recognised, respected and fostered in the formal teaching of geography. At the academic level there is a case that the existence of private geographies sets new grounds for the epistemological foundations of geography and the function of geographic education. Spatial experiences and feelings are an integral part of our living (Relph, 1976), just as locational analysis is part of our attempt to understand the structure and organisation of space.

To Relph the elements which give identity to a place are: its physical setting, the activities common to the place; and the meanings invested in the place. These three attributes may be subsumed within the general meaning of spirit of place, sense of place, or genius of place—not a clear-cut concept to put across in a classroom. Judith Briscall (1980) has suggested that while it is impossible to teach students how to perceive the uniqueness of a place, they may be introduced to the idea and reality of a sense of place. The following procedure is suggested as an activity for middle and upper forms. Direct the following questions to students who have been willing to bring in a slide of a place they feel strongly about: 'what do you like/dislike about it? In what way is the place shown impressive or depressing, exciting or dull? How did/do you feel there—at home, frightened, calm, restless? What memories does it have for you? Is there anything about the place which is not captured by the camera?'

From attempts to respond to these questions, Judith Briscall suggests that a discussion may develop on the idea of a 'spirit of place'. General ideas to be cultivated include the following: that sensitivity to the atmosphere of a place varies from person to person (and from culture to culture); and that the spirit of place can be intentionally created. For example, we may ask to what extent can architects create atmosphere through design and building materials? To what extent can we create atmosphere by personal arrangement and decoration? What comparisons could be made between the intentional 'place-making' of eighteenth century landscaped gardens and modern New Town designs, for example?

The personal geography of early settlers

We may also go back to the American HSGP example used earlier in the chapter and consider it in conjunction with a model sketched in Figure 1.4. In his model, R. Simson (1976) identifies some of the possible sets of ideas, beliefs and images which may have informed the environmental perceptions and behaviour of early settlers in the river valleys of eastern Australia. Given his scenario where might you decide to settle and why? How would you choose to earn a livelihood? How would you affect the environment? Simson suggests that settlers 'set about transforming the environment in a deliberate but unplanned way until its ecological stability had been lost and its environmental character drastically altered'. It is interesting to compare the appraisals accorded to early settlers with that of a present-day conservationist who has a very different set of attitudes, and goals. Simson sets these out as in Figure 1.5 where eroded slopes and derelict farms tell a story and in turn influence present attitudes and values. The potential for using these models in a values analysis exercise (see Chapter 4) is considerable. Here, they stand as scenarios exposing the influence of perceptions, attitudes, values, needs and beliefs on settlement decision-making and quite clearly they may be seen as a possible variation on the earlier settlement siting activity.

Linking science and experience

An exploration of private geographies and an appreciation of their foundations in personal experiences, preferences, attitudes and feelings may give students a sense of how geography as locational analysis may

Figure 1.4 Possible elements in the personal geographies of early settlers in assessing the river valleys of eastern Australia

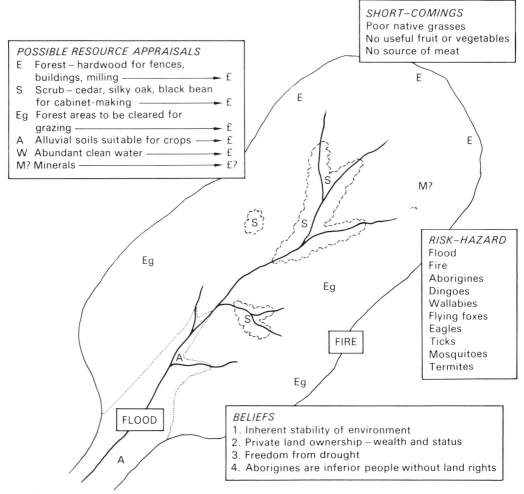

POSSIBLE RESOURCE APPRAISALS
E Forest – hardwood for fences,
   buildings, milling ————————→ £
S Scrub – cedar, silky oak, black bean
   for cabinet-making ———————→ £
Eg Forest areas to be cleared for
   grazing ———————————————→ £
A Alluvial soils suitable for crops ——→ £ £
W Abundant clean water ——————→ £
M? Minerals ———————————————→ £?

SHORT–COMINGS
Poor native grasses
No useful fruit or vegetables
No source of meat

RISK–HAZARD
Flood
Fire
Aborigines
Dingoes
Wallabies
Flying foxes
Eagles
Ticks
Mosquitoes
Termites

FIRE

FLOOD

BELIEFS
1. Inherent stability of environment
2. Private land ownership – wealth and status
3. Freedom from drought
4. Aborigines are inferior people without land rights

Source: Rob Simson, Principal, Maroon Outdoor Education Centre, 1976.

be enriched. It may be enriched by bringing the dimension of felt experience to the surface. This experiential base to intellectual decision-making is poetically described in the following account of how a Japanese architect consciously plans to experience a place before he applies his rational knowledge of the design process:

Quite simply, if designing say, a residence, I go each day to the piece of land on which it is to be constructed. Sometimes for long hours with a mat and tea. Sometimes in the busy part of the day when the streets are a bustle and the sun is clear and bright. Sometimes in the snow and even in the rain, for much can be learned of a piece of ground by watching the rain play across it as the run-off takes its course in rivulets among the natural drainage ways.... I go to the land, and stay, until I have come to know it. I learn to know its bad features.... I learn to know its good features....

And so I come to understand this bit of land, its moods, its limitations, its possibilities. Only now can I take my ink and brush in hand and start to draw my plans. But, strangely in my mind the structure by now is fully planned, planned unconsciously, but complete in every detail. It has taken its form and character from the site and the passing street and the fragment of rock and the wafting breeze and the arching sun and the sound of the falls and the distant view.

Knowing the owner and his family and the things they like and the life they would like best to lead, I have found for them here on this land the pattern of living that brings them into the most ideal relationship with their land and the space around them, with their living environment. This structure, this house that I have planned, is no more than an arrangement of spaces, open and closed, accommodating and expressing in stone, timber and rice paper a delightful pattern for their life on this land. How else can one plan the best home for this site?[5]

Figure 1.5 Possible elements in the personal geographies of a conservationist assessing the river valleys of eastern Australia

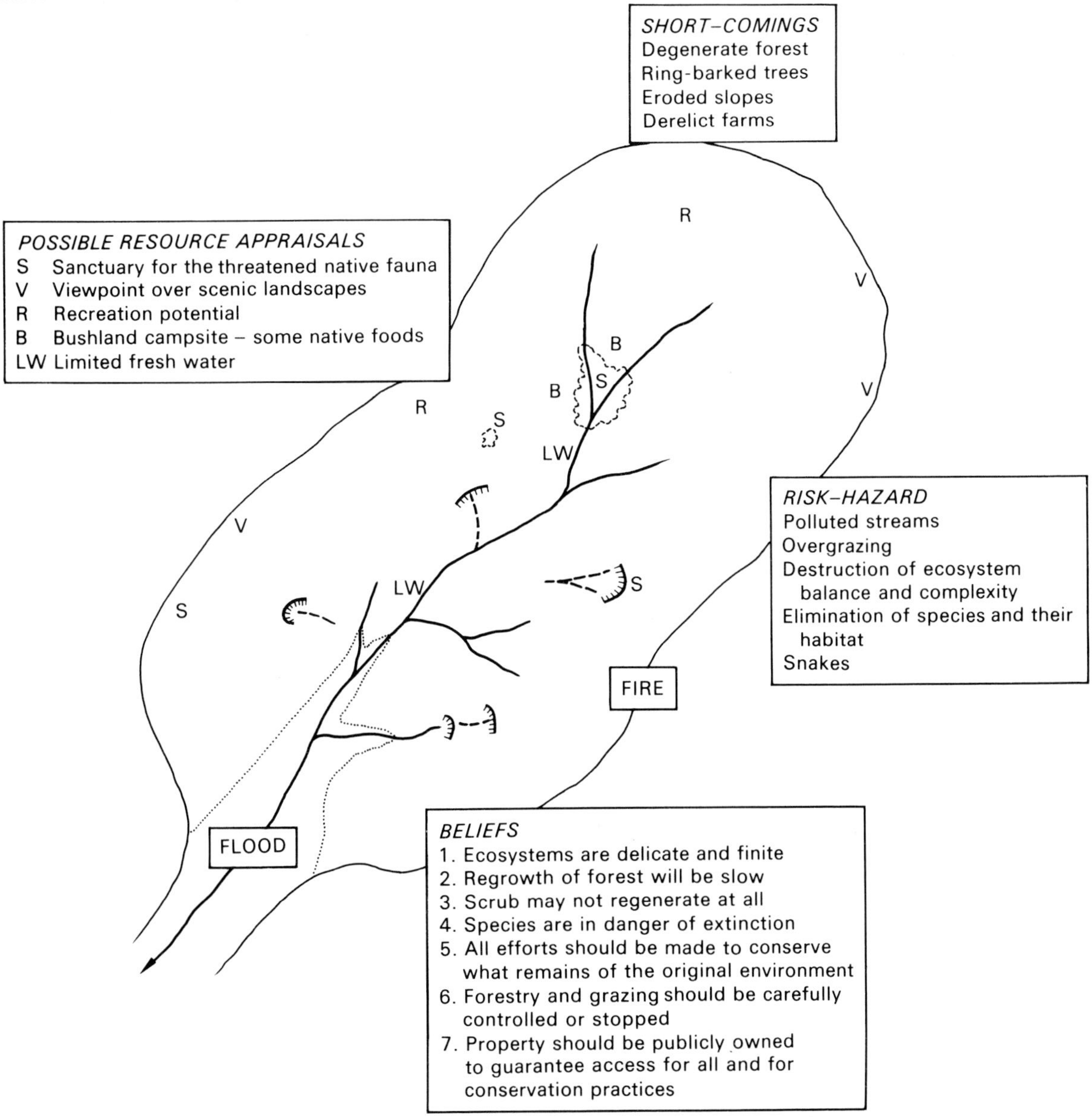

SHORT–COMINGS
Degenerate forest
Ring-barked trees
Eroded slopes
Derelict farms

POSSIBLE RESOURCE APPRAISALS
S Sanctuary for the threatened native fauna
V Viewpoint over scenic landscapes
R Recreation potential
B Bushland campsite – some native foods
LW Limited fresh water

RISK–HAZARD
Polluted streams
Overgrazing
Destruction of ecosystem
  balance and complexity
Elimination of species and their
  habitat
Snakes

FIRE

FLOOD

BELIEFS
1. Ecosystems are delicate and finite
2. Regrowth of forest will be slow
3. Scrub may not regenerate at all
4. Species are in danger of extinction
5. All efforts should be made to conserve
   what remains of the original environment
6. Forestry and grazing should be carefully
   controlled or stopped
7. Property should be publicly owned
   to guarantee access for all and for
   conservation practices

SOURCE: Rob Simson, Principal, Maroon Outdoor Education Centre, 1976.

Knowing and feeling

Geography as science and geography as personal response are not to be seen as opposing views of the discipline. The element of known attitudes and values in understanding and decision-making has been acknowledged in recent explanations derived from behavioural geography. Such considerations are now beginning to appear in textbooks and geography teaching journals. It seems clear that more humanistic questions should be included in learning activities in order to enhance students' everyday ways of organising spatial experiences. Once refined and expanded, these may then become conscious rather than unconscious dimensions of thinking and action. Perhaps, the interplay between private geographies

Figure 1.6 Steeple-chasing in Oadby—an experience of serial vision

STEEPLECHASING IN OADBY
AN EXPERIENCE OF SERIAL VISION

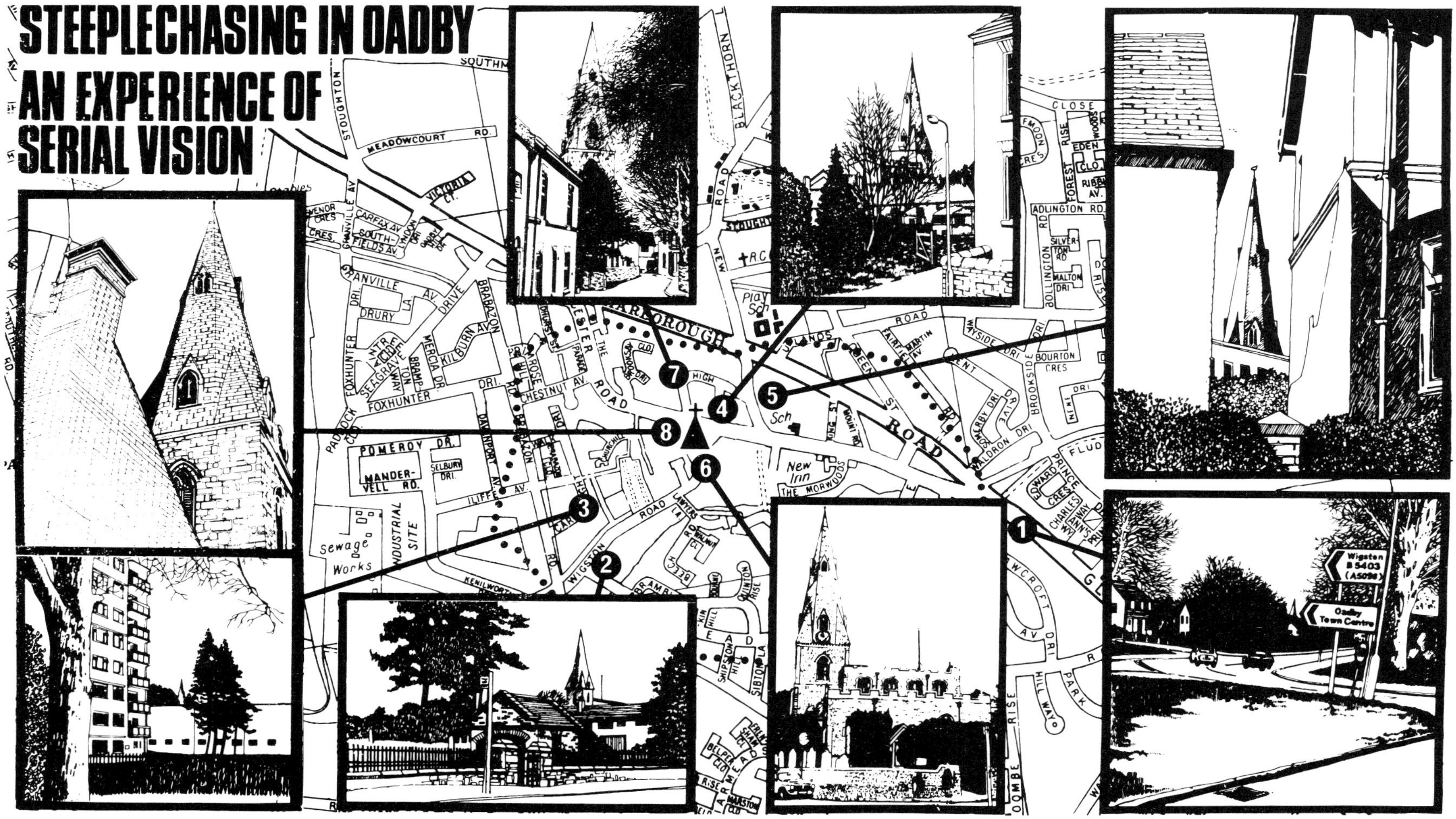

SOURCE: Wheeler, Keith, 'Assessing Townscape', Schools Council Art and the Built Environment Project. See also *Bulletin of Environmental Education*, No. 102, p. 18.

and 'formal' geography, will enable students both to clarify their feelings and attitudes about their environments, and to realise the reciprocal nature of the impact their *lived-in* space exerts on them and they on it (see Ward and Fyson, 1972).

Those who feel uncertain about the educational necessity of more humanistic approaches in geographical education might like to consider Kevin Lynch's *Growing up in Cities* (1977). Lynch's work is an empirical investigation of the way small groups of teenagers use and value their environments. Much is revealed of the relationship of young adolescents to their environment. The reports are full of human detail, local colour, illuminating fact and vivid impression. The investigation procedures included observation, interview, recording time budgets and sketching mental maps. The questions put to the teenagers in interview included:

1. As you go about your usual day's activities, what particular places or things give you the most difficulty? ... Are there places you can't get into and wish you could?
2. Do you help maintain or fix up any part of your area? ... Are there any places that nobody owns?
3. Where do you best like to be? ... Where is the best place to be alone?
4. Has your area changed in your memory? Do you think it has become better or worse?
5. On what occasions do you go out of your own area?
6. Are there beautiful places in the city? Why are they beautiful?
7. Of all the places that you have ever been in, or heard about, or imagined, what would be the best place to live in/the worst place?

Questions like these offer opportunities for further exploration and elaboration in field work, for example, and could be used to plan a sequence of work. A study of beautiful or ugly places may require a photographic record or field sketches of various places. Wheeler's strategies (1976) of steeplechasing represented in Figure 1.6, and townscape notation are useful as exercises in aesthetic education. Analysis of the places may also be by form, function, size, building material, location, accessibility, frequency of use/visit. This kind of analysis is suggested not so much in the spirit of scientific endeavour as to encourage

students to understand the sort of things, including their own activities, that contribute to and influence their perception and interpretation of environmental experience.

The questions asked by Lynch deserve further elaboration and some incorporation into the spirit and purposes of learning through geography in schools. The answers he received from students suggest that they have *strong* ties with their environment but *limited* conceptions of its potentialities. Significantly, their ideas of their possible roles in environmental-planning and decision-making are also limited. The evidence amassed suggests that children need more opportunities to explore and, both cognitively and affectively, evaluate their environment. Traffic survey and frequency counts are not designed to promote strong feelings of individual or collective engagement with environments. Personal dimensions of environmental-knowing need encouragement, analysis and evaluation.

Conclusion

The learning activities described in this chapter have all been planned around an initial question or questions. The case for structuring activities around guiding questions and related sub-sets of questions has been based on the idea that the process of generating meaningful learning and understanding is fundamentally tied to demonstrating quite explicitly the link between questions and 'answers'. 'Answers' mean nothing if we do not know the questions to which they are intended to be a response. If this is so, then it is logical to identify questions as the initial step in activity planning. Evidence suggests that the choice of questions and the planning of their presentation for analysis is often most easily and usefully achieved by a small group of teachers brainstorming and subsequently refining and structuring their ideas into balanced courses. Just as maximum open-mindedness is advocated in the initial stages of activity planning, the questions cited in this chapter were enlarged upon in the activity outlines to encourage a movement towards open-ended, speculative responses.

Question identification can be usefully adopted as the first procedure in planning a learning activity. Questions are thus the initial and continually guiding

signposts which help us to organise and plan path-
ways leading students to meaningful learning
through geography.

Further reading

1. *SGEP-PAK* (1977) contains many examples of enquiry planned teaching organised around the question identification approach.

2. More readily available are the textbook series,
Going Places, consisting of three books: *Out and
About*, *Way Out* and *Out of Site*. These textbooks,
especially the first two, contain learning activities
clearly organised around key questions and each
chapter is an activity or group of activities rather
than a chapter in the conventional sense. Listed
below are the contents of each book to illustrate
more fully the question posing orientation.

Out and About (Blachford, Brown, May, Rickards
and Wellard, Rigby, 1975)
1. How well placed is the school?
2. What are the effects of a flood?

3. Where does the mail come from?
4. Is the desert a sea of sand?
5. Kangaroo—purse or pouch?

Way Out (Blachford, Brown, May, Rickards and
Wellard, Rigby, 1976)
1. Does your school have a traffic problem?
2. Shake, rattle and roll.
3. What is your neighbourhood?
4. How do cities get water?
5. Are there regions of disease?

Out of Site (Wellard, May, Hartnell and Brown,
Rigby, 1978)
1. The where and why of urban places.
2. Who's in cyclone country?
3. Where are the best places?
4. Where are all the good things made?
5. National parks—the question of preservation.

3. For ideas readily translatable into lesson plans of
a humanistic kind consult *People in Places* (Farbstein
and Kantrowitz, Prentice-Hall, 1978).

2 Planning Learning Activities to Reach Generalisations and Decisions

Planning for generalisations

In his telling analysis of two views of the teaching/learning process, summarised in Figure 2.1, Nicholas Helburn (1968) drew attention to the importance of developing generalisations and broad understandings in the process of learning through geography.

This chapter is organised around the notion that the process of teaching and learning should lead towards an enriching of general concepts, or possible general answers, and general ideas, or towards reaching decisions.

A significant element of planning activities needs to be given over to organising procedures and strategies for assisting the process of connecting ideas to, and focusing ideas and concepts around, a yet more generalised idea or understanding. By identifying questions and building links to reach generalisations and decisions we are going some way towards planning lesson activities which will stimulate thinking about relationships. When little knots of understanding are tied into larger knots, sense and learning coalesce.

Where of course we are aware that the intellectual maturity of students is likely to predispose them to

Figure 2.1 Two views of the teaching/learning process

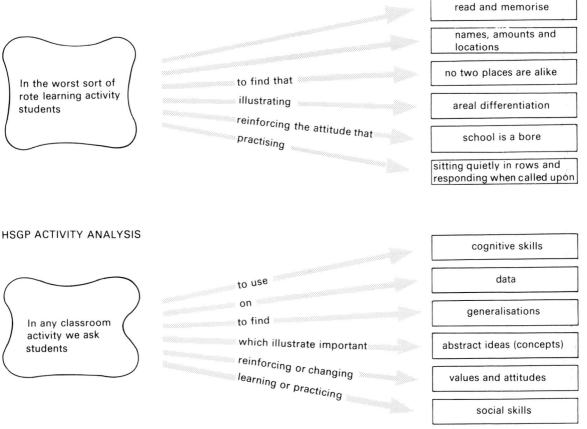

ROTE LEARNING ACTIVITY ANALYSIS

In the worst sort of rote learning activity students

to find that
illustrating
reinforcing the attitude that
practising

read and memorise

names, amounts and locations

no two places are alike

areal differentiation

school is a bore

sitting quietly in rows and responding when called upon

HSGP ACTIVITY ANALYSIS

In any classroom activity we ask students

to use
on
to find
which illustrate important
reinforcing or changing
learning or practicing

cognitive skills

data

generalisations

abstract ideas (concepts)

values and attitudes

social skills

Source: Helburn, N. (1968) 'The educational objectives of high school geography', *Journal of Geography*, Vol. 67, No. 5, National Council for Geographic Education.

syncretic reasoning rather than logical thought then experiences and resources of an enactive (action based) and an iconic or image, graphic type presentation will be more appropriate than symbolic representations of data (Bruner, 1966). Long and Roberson (1968), and indeed many others, emphasise the use of a wide range of visual materials in geography teaching at all levels but such a commitment is crucial at junior levels.

The focus on planning for developing understanding of generalisations and encouraging decision making is central to the message of this chapter. It links logically to the question–answer argument of the first. This message will be expanded after a brief consideration of the place and positioning of objectives within planning.

Objectives—specific and general

Defining and making our purpose explicit is admittedly not an easy task no matter when it is undertaken and some discussion is necessary. The chief difficulty very often lies in stating objectives at a level of specificity which enables one to have a clear idea of (1) what will be taking place in the classroom, of (2) knowing what the students will be doing and learning, and (3) what understandings they will be working towards.

Specific tasks

The difficulty of an appropriate level of specificity (see Marsden, 1976) can be overcome if we set objectives or tasks at two levels of precision, the specific and the more general. For example, if students are counting and placing in a matrix the number of road links joining a set of places, then they are engaged in those very specific tasks or activities. In addition, it can be assumed that a teacher has set those tasks with a view to answering the question, 'What are the most and least accessible places?' so as to develop or reinforce the concept of accessibility and the idea of different places having differing degrees of accessibility. The tasks of counting road links and building a matrix are very specific objectives or tasks. These tasks are necessary prerequisites to reaching the more general objective or task of understanding accessibility as measured by road links.

Clearly, specific objectives may be equated with or seen as a mirror image of the tasks the students have to accomplish. Descriptions of planned student activities can stand as specific objectives, although some may object to such equating of doing with learning. If it is accepted that activities mirror specific objectives, then especially in the apprenticeship days of activity planning, it is practical to move straight from identifying the question sequence to sorting out what activities and details of organisation are needed to follow up the questions.

Cognitive styles

Planning work may proceed more readily if one moves from the key questions to their translation into an activity, rather than if one is asked to state a number of related objectives and then decide on how to achieve them. I prefer, as Figure 2.2 indicates, to leave the identification of objectives to a much later stage in lesson planning.

To encapsulate specific objectives and learning activities may be a mildly heretical position given the present dominance of the objectives model in the literature of curriculum planning in geography. Perhaps one can bow out of any argument by suggesting that the rationale for planning depends on one's cognitive style and preference or even one's teaching style and preference. Eisner (1979) has recently linked views on planning with teaching style. If this is so, then we must proceed as best suits our cognitive or teaching style, acknowledging nevertheless that concepts from educational theory such as objectives may inform practice and shape planning directly and indirectly. It may be that when specific learning activities have been listed, we should then ask ourselves, 'What is being achieved through these activities? Do they contribute towards answering the questions in the enquiry sequence? Why are we dealing with these questions?'

General objectives and generalisations

Stating specific objectives and detailing learning activities have thus been collapsed into one, though there may well be a need to evaluate the activity for its appropriateness at a later stage and to assess what purposes it is likely to fulfil.

Figure 2.2 Key steps in planning activities

1. Brainstorm for questions.

2. Cull the list for the 'best' questions.
Are the questions important?
Are the questions geographic?
Are the questions likely to be motivating to the learner?

3. Define sets of sub-questions (an enquiry sequence) appropriate to each of the key questions.

4. List the concepts, generalisations, central understandings you consider you are planning for.

5. Brainstorm for appropriate student activities and teaching strategies—give special consideration to ideas for initiating the activity.

6. Consider resources and materials—consider what already *exists* and what could be *developed*. What data bases are appropriate?
How could the information be explained and presented?
In what order?

7. Select the most appropriate student activities and teaching strategies. Be aware that student tasks equate with specific objectives.
Be aware that student tasks are a means of reaching generalisations. Is there a balance and range of tasks?

8. Decide on the form and organisation of the tasks i.e. what data and what data processing methods will be used? In what order?

9. Consider the objectives arising from the questions at least in general terms and in the light of the general ideas to be developed.

10. Develop assessment and evaluation procedures. Match these to key questions and activities.
Consider the range available, formal, informal, written, verbal. What congruence, what contingency, within and between activities?

SOURCE: based on Secondary Geography Education Project (1977) *SGEP-PAK*, Geography Teachers Association of Victoria.

What of general objectives? Let us return to the idea of a mirror image again. The general objective may be thought of as the mirror image, the equivalent of the generalisation or principle, the general notion or idea which is being worked towards. The general idea will match up with the answers to the questions in the enquiry sequence. Hence the way in which, and the level at which, the questions and sub-sets of questions are posed will determine the kind and level of generalisation to be taught and stand in as our general objective. A generalisation is therefore equivalent to a general objective, the end to which learning activities have been planned and organised. Pring's point (1973) that our objectives may be transformed as we respond to the dynamics of the teaching milieux we find ourselves in is nevertheless valid but we have to begin somewhere with some question or goal in view.

Planning lessons to reach generalisations centres around arranging resources and tasks to work towards answers to questions. Activities should move towards the articulation of general statements or decisions which provide students with some kind of overview and resolution, however limited or tentative, of the *meaning* of the work they have been doing.

Going beyond the information given

The generalisation or decision represents, in Bruner's words, where 'going beyond the information given', takes one. In reaching an understanding of the meaning of the general point, in Piaget's broadest description of the learning process, students will be assimilating and accommodating new ideas and thinking to already existing cognitive structures. It goes without saying that as communication and exchange among teachers and students increases, teachers develop a sensitivity towards, and an intuitive feel for, how best to arrange resources and activities to facilitate the assimilation/accommodation process and to promote the ability to go beyond the information given.

Planning at the instructional level—selecting questions

It has been assumed in the kinds of questions and activities suggested in the first chapter that they fit in with an unspecified overall programme which shall remain unspecified since we are more concerned here with activity planning than course construction.

Biddle (1976a) and Graves (1979) would use the phrase, planning at the 'instructional' level, to refer to planning a sequence of lessons in contrast to the level at which programme or course planning takes place. 'Activity' seems a preferable term with its sense of people involved in doing and interacting rather than the idea of people being instructed, necessary though that is as one of the activities appropriate to learning through geography. We should recall that activity and interaction, i.e. interaction in the form of discussion and co-operation is one of the factors that promote intellectual development. In assuming the appropriateness of the questions and activities suggested to an overall programme, the issue of (1) how questions are selected for, or culled from, lists or (2) what makes them suitable questions in an overall sense has only briefly been raised.

Perhaps a little more expansion of this issue is required. SGEP suggests that the following questions may be applied to evaluate a range of suggestions. These may be applied just as readily to an enquiry sequence:

1. Is the issue raised significant?
2. Does it raise issues students will probably need to consider now and in the future?
3. Is the problem or question geographic?
4. Is the problem or question appropriate to the students for whom it is intended?
5. Is the issue, question or problem likely to motivate interest and application?
6. Does it connect with student experience?
7. Is there a balance of topics, areas and types of questions?
8. Is there a balance and variety of specific and general purposes?

These questions are useful and may be related back to ideas in curriculum theory which suggest that a balanced course in any subject should take into account the nature of the subject, the needs of the students, and the wider society of which they are part.

The nature of the subject—choosing a paradigm

The notion of geography's paradigms introduced into geographical education by Biddle (1976b) are helpful in planning, and they are also useful in evaluating enquiry sequences. They lead one to ask: What view of geography is being taught? Is this the view desired? Should other views be included? Do we want some explanation of regions and landscapes as well

as spatial systems and environmental experiences? Evaluating enquiry sequences for their geographical perspectives is important as obviously students can be exposed to a number of perspectives if we wish. 'Is the problem geographic?' has more than one meaning and this needs to be taken into consideration. It has been decided in this book to include both scientific and humanistic views to illustrate their complementarity and the value of both to environmental understanding and decision making.

Choosing topics

The needs of the student and the wider society have been closely examined in all geography projects and rationales relating to these questions may be consulted in the documentation put out by the various projects. Here we shall confine ourselves to three points. It is necessary to motivate and interest students and so topics of immediate interest may be chosen for development into learning activities. We also need to remember that it is important to try to interest students in questions likely to be of future significance, so that present learning has future value. The relevance of geography to understanding society and contributing to the student's participation in that wider society beyond school and its requirements demands a mix of issues and a mix of these issues should be spread across local and national arenas into international society. What environmental, planning or First World/Third World issues need to be raised? What issues relating to social justice should be explored? What issues relating to human justice at an international scale should be explored?

Structuring and restructuring generalisations

Let us now return, after that brief excursion into the specific position of objectives in the planning procedure, to the main message of the chapter. The emphasis on structuring activities to reach generalisations has a number of dimensions. While, initially, working towards generalisations as such will be emphasised, other related ideas should also be appreciated. These are the notions that some teaching will be directed towards the following ideas: that it is often necessary to modify or restructure general ideas (and teaching strategies for this purpose are available) over and beyond modifications which may occur in classroom dynamics; and that generalisations are often used and should be used in subsequent thinking (this occurs most obviously in planning and other hypothesis-generating and decision-making exercises).

Concepts and generalisations as advance organisers

As an understanding of concepts may be seen as acting as advance organisers in the building up of general ideas (Ausubel, 1960) so general ideas may be used to move thinking on, to restructure it or inform decision-making.

The teaching ideas described in this chapter progress through this kind of hierarchy. General ideas are developed in order to create meaningful connections between facts and ideas. The general ideas are then applied in other ways so that understanding is refined and elaborated or used to further refine problem-solving and/or inform decision-making. It should be noted that this suggestion of a hierarchy does not imply a hierarchy of difficulty but one of function or purpose.

It is clear that the categories of concepts, generalisations and decisions possess within category variation in difficulty, both in relation to the concepts themselves and the examples. Quite simple concepts and very abstract concepts fall within the category of concept and likewise quite simple and quite complex ideas may be modified, or restructured in the process of decision-making and synthesis. Further variation in difficulty obtains in the sample activities outlined in this chapter because they have been chosen to illustrate both geography as science and geography as personal response.

Classroom activities

Moving towards generalisations

I shall now attempt to suggest ideas for lesson activities with a view to emphasising the general notions and relationships which may emerge. In Chapter 1 we touched on the kinds of generalisations the settlement siting activity leads to—for example, the change in variables influencing location as transport technology develops (or the changing influences of cultural variables in England's long history of settlement development).

An enquiry sequence which continues the settlement theme includes the following questions:

Are there *patterns* to be found in the distribution of settlements?
Is there a relationship between the size and *spacing* of settlements?
What relationship seems to exist between the size of settlements and their number?
What relationship exists between the size of settlements and the *distance* between them and their neighbours?
What *relationship* exists between the size of settlements and the number and types of communication linking them?
What pattern is there to the kind and number of *links* between settlements of the same size?
What *pattern* is there to the kind and number of links between settlements of different sizes?

The map of the United States urban patterns (Figure 2.3) is intended to serve as an introduction to the activity planned for finding answers to this question sequence. It relates to the first question 'Are there *patterns* to be found in the *distribution* of settlements?'

Describing patterns

A number of adjectives could be used to describe and classify the patterns in Figure 2.3. In Nevada, settlements are very spread out or dispersed and there are few of them, while in New York and adjoining states there are many settlements which are very large and clustered together. By contrast, the Iowa pattern is a fairly even one with a greater number or density of settlements than in Nevada and the settlements are closer together. Idaho and Utah provide a contrast: a linear pattern of settlements related in fact to the irrigated farmland at the edge of the mountain ranges.

An examination of a map of settlements in the United Kingdom would establish a similar variety in settlement arrangement or patterns, density and dispersion. Clustered conurbations occur around coal fields and ports for instance. Some discussion of the relationship of settlement distribution to resources and transport connections might be useful if it were to relate back to the factors influencing the development of settlements. The activity is intended here however, to be brief and introductory, to orient students to the idea of how settlements are distributed.

Figure 2.3 United States: urban patterns

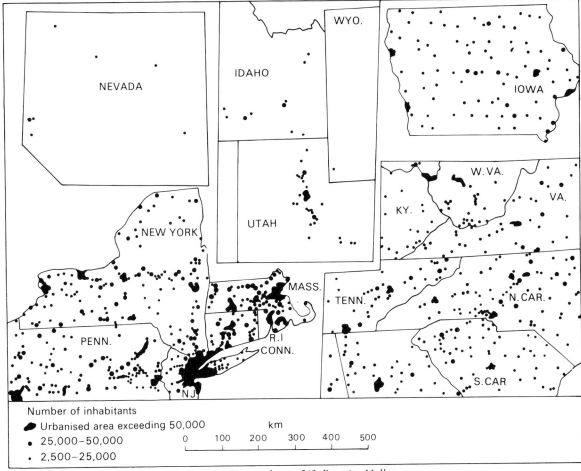

Number of inhabitants
- Urbanised area exceeding 50,000
- 25,000–50,000
- 2,500–25,000

km
0   100   200   300   400   500

Source: Alexander, J. W. (1963) *Economic Geography*, p. 562, Prentice-Hall.

Patterns of settlements

A topographical map (Sheet 7925 (Edition 1) Series R652) of the Shepparton district, a rural area in Victoria, Australia, has a distribution of settlements not unlike that of the larger scale Iowa map. Settlement can be seen to be fairly evenly distributed and the towns and cities are market places where people mainly buy and sell goods and services. There are no great port settlements, industrial cities or mining centres here. The Shepparton map may be used to examine the patterns and relationships between settlement and communications as an attempt to answer the remaining questions outlined earlier and reach general understandings based on relationships between the two features. Only a sketch map showing general relationships is reproduced here in Figure 2.4.

Classroom questions

The questions a teacher might actually use with a class to examine the patterns of settlements and communication on the topographical sheet could include:

1. What appears to be the largest settlement? Why do you say this? How many and what kinds of transportation routes serve Shepparton? How does this compare with other settlements?
2. Name settlements which are shaded but are smaller than Shepparton. On what kinds of roads are they located? What connection do these towns have with Shepparton? On average, how far are the smaller towns from Shepparton? On average, how far are they from each other?
3. What other settlements are shown on the map? What buildings do some possess and others not? On what kind of roads are these settlements

Figure 2.4 Settlements in the Shepparton area

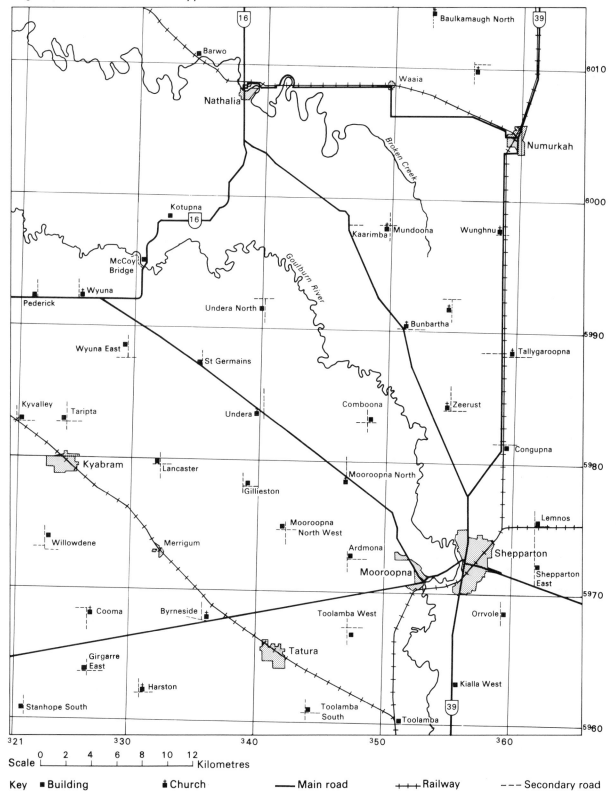

Scale

Key ■ Building   ♟ Church   —— Main road   +++ Railway   – – – Secondary road

located? How far are these places from their nearest neighbours of the same size?

4. Are there any other indications of settlement? How widely do they occur? What roads serve them?

The largest town on the map, by area, is Shepparton. With a class it is helpful to construct a grid—on the blackboard or on an overhead transparency, and to shade in Shepparton and other features mentioned.

It is the amount of shading—the built up area—which indicates Shepparton's importance. It appears to have, by implication, the most buildings and presumably the largest population. As well, Shepparton also has a greater number of lines of communication running through it than any other settlement. There are two principal highways, a number of secondary roads and a single track railway converging on the town. No other town has an airport.

Let us examine seven smaller built up areas, Mooroopna which neighbours Shepparton to the west, Tatuna to the south-west, Kyabram and Merrigum due west and Nathalia, Waaia and Numurkah to the north. What connections do these smaller towns have with Shepparton? What kinds of roads link the towns to Shepparton, the largest settlement? Reading the topographical map reveals that all of the smaller towns are connected by rail to Shepparton and three of the seven are located on a principal highway. None, unlike Shepparton, is situated at the focus of two principal highways though three, Nathalia, Numurkah and Mooroopna, are on the crossroads of a principal and a secondary road. Three others are on secondary road intersections while only one of the smallest places, Merrigum, lies at the crossing of minor roads. Two towns are on principal highways directly linked with Shepparton; the remainder, apart from Merrigum, all have access via secondary roads. Quick sketch maps of the towns, in relation to the type of highway running through them, would illustrate the contrast with Shepparton. On average, the settlements are, as the crow flies, about six kilometres from their nearest neighbour, and about twenty-eight kilometres from Shepparton.

Two sizes of settlement, their number and distribution, together with their transport connections have been examined. Other settlements are also shown on the map. Altogether, another thirty-seven places are named. With junior classes, it is advisable at this stage to confine the counting of settlements and their position on routes to one half or even one

quarter of the topographical map if this is available or the exercise becomes too long and demanding of concentration. The position of places on communication lines can be labelled according to the highest order route serving them. A place at the crossroads of a main road and a secondary road for example is classified as being on a main road. Just less than one quarter of all places are either on a main road and railway, or a main road only, or a railway only. One eighth are on secondary roads. Two thirds are on minor roads and distances between the smaller places average four and a half to five kilometres. Of the isolated unnamed buildings many of which are farmhouses, it is estimated that over eight hundred are scattered over the map, the majority being closest to minor roads.

Generalisations which emerge

Let us now go through the enquiry sequence to indicate what general points emerge.

1. What relationship seems to exist between the size of settlements and their number? *The larger the settlement, the fewer they are.*

2. What relationship exists between the size of settlements and the distance between them and their neighbours? *The larger they are, the further apart they are.*

3. What relationship exists between the size of settlements and the number and types of communication linking them? *The larger they are, the greater the number of links and types of connections with other places.*

4. What pattern is there to the kind and number of links between settlements of the same size and of different sizes? *The smaller settlements are usually linked to one another by minor roads, which in turn join secondary roads leading to larger settlements and principal roads and railways. Settlements are usually linked to one another by the same type of road and then to the next largest settlement by a higher grade of road.*

Analysis of settlement and communication patterns establishes several general principles of settlement hierarchies and central place theory—without necessarily using the term or introducing the associated geometry.

This activity was originally planned by Alison Doggett (see Slater and Spicer, 1981) for use with a class of twelve- to thirteen-year-old English suburban children in examining the O.S. metric extract 149M/ 92 of Malton and Norton. A double-period lesson progressed to the stage where children articulated for themselves the general statements suggested here. Indeed, this seemed evidence that anything can be taught to any child at any stage of development in some intellectually honest form. The lesson was a positive example of the high levels of thinking which may be achieved when Bruner's more open philosophy of courteous translation replaces the rigid interpretation of mental development derived from Piaget. While students in this case worked towards these generalisations it is granted that there will always be exceptions to generalisations, since towns and roads are features of the human built environment and therefore based on human decisions. Generalisations, explanations and predictions of locations and patterns can at best be phrased in probabilistic languages.

Relationships between two variables have been examined and clarified. In order to work through the questions, it is necessary for a student to be able to identify and define from information on the map, settlements of different sizes and types of communications. In the process of identification and definition, a description and classification of these two elements emerges. It is the application of these four intellectual skills which then permits an analysis of relationships between settlement and communication. *What* is related and *how* they are related has been established.

Classroom activity

Mapping shopping habits

The previous question sequence did not encompass an explanation of the settlement hierarchy. We could move into another enquiry sequence and set of lesson activities by asking:

1. Where do you, or would you or your family most frequently buy: (a) bread (b) other food items (c) ordinary everyday clothes (d) rather special furniture (e) an original painting.
2. For which goods do you have the greatest number of shops to choose from?
3. How much money (small, medium, large amounts) do you and your family spend *per visit* on the different items?

4. How many visits are made per week, month and year to the different types of shops?
5. What differences exist among the trade areas for the items?
6. What relationship seems to exist between the item purchased and the cost and frequency of purchase?
7. Why are there settlements of different sizes?

A mapping exercise is an appropriate way in to establishing evidence from which the answers might be reached. Four or five maps of the anticipated reach of shopping travel are needed. Each should be covered with tracing paper. One map will be used to build up each purchasing pattern. Before using the maps, each student should make a list of the places patronised and then mark in the location of home by a dot and rule a straight line from home to the shops where the goods are bought.

As the completed maps are studied in turn and the number of places patronised on each map counted, it will be observed that there are a larger number of food shops, and that they are closer to one another than, for example, furniture shops. Class discussion should establish that the number of sales a shop needs to attract in order to make a profit and stay open is the threshold for that good. Lots of quite small purchases made often, of the kinds of things we need often, e.g. bread, can support more of the kind of shops selling those goods. Where people buy something more expensive, probably only once a year or even less, e.g. furniture, then the number of shops selling those goods will be fewer and people will travel further from home to reach them. The distance people are prepared to travel for daily needs is less than for more expensive, less frequently needed goods. The range of a good differs from one type of good to another.

A comparison of the maps will point up the usefulness of the concepts of threshold and range. The trade areas or hinterlands for each good will be different, the variations in the size and overlap of trade areas illustrating the range of a good. The following generalisations, amongst others, generally emerge:

1. People usually travel only short distances for everyday items but longer distances for goods not needed so often.
2. The greatest choice of shops exists for everyday goods.
3. Less money is spent per visit at the shops visited most frequently.
4. More visits are made to shops selling goods needed often than to those selling goods needed less often.

5. The trade area is smallest for the shops visited most often, and largest for the goods bought least often.
6. The less costly goods are the more frequently purchased goods and the most costly goods are bought much less often.
7. Since some goods are needed more often than others, many small towns and neighbourhood centres with just a few shops grow up to supply these goods. A few small towns and neighbourhood centres develop into larger towns and centres as shops selling less frequently purchased goods are set up and attract customers.

These two learning activities concerning settlement size and spacing and shopping habits explore relationships and establish generalisations framing those relationships. The connection between questions and answers is clear. (Readers may like to refer to Bennetts in Walford (ed.), *Signposts for Geography Teaching*, for an elaboration of progression in geography lessons. A set of examples is specifically related to settlements and shopping habits.)

Challenging generalisations through hypothesis testing

At this point, the reader may feel that the emphasis placed upon reaching generalisations in lesson activities has been unrelentingly pursued. At the same time it has been assumed that the data—the evidence—presented has provided learners with both necessary and sufficient means to reach accurate conclusions, conclusions which would enable a reasonable prediction to be made about the relationships among a set of variables. Forming general notions provides us with something on which to base our expectations and hypothesis-testing is now a well-established method in geography teaching for reviewing generalisations. Patterns of shopping and consumer behaviour can be tested, using data held in computers.

In two recent reports (Shepherd, *et al* 1980 and Stevens, 1980) there are accounts of computer work examining shopping patterns and testing hypotheses. The computer programme GRAVITY may be used to test consumer shopping habits between centres. The programme is based on Huff's model or generalisation and it estimates the probability of a consumer

visiting a centre by taking into account the attractions of all potential shopping centres simultaneously. A student used the programme to test shopping choice in the St Albans area of England. The probability of shopping in each of four centres was calculated by GRAVITY. The model predictions were then compared with fieldwork results. The differences emerging were noted and an analysis of possible reasons made. The generalisation embodied in the model—namely that the probability of a consumer travelling to a centre is dependent on the size of the centre, the consumer's distance from the centre and willingness to travel varying distances for different goods and services and the competing attractiveness of all centres—can be modified, enlarged and re-established. It may be stated that generalisations are not for believing but for testing and the reformulation of generalisations is a constant process in learning through geography. Teaching strategies for challenging generalisations are often practised in a hypothesis-testing manner.

Discrepant data procedures

An exciting and under-used elaboration on the 'teaching for general understanding' theme is the discrepant data procedure. Like hypothesis testing it also enables generalisations to be modified, tested, and reinterpreted. The American HSGP pioneered this technique (HSGP, Unit 1, 1969). Essentially the discrepant data procedure is designed to have students examine a generally held idea or belief. An activity is structured to reinforce the idea and then new evidence is introduced, which challenges the idea and leads to a restructuring of the generalisation. In other words, the discrepant data procedure is a strategy for promoting more precise and accurate general notions.

Cultural diversity

For example, it is likely that most students have high expectations of being able to identify and accurately assign half a dozen slides of cities from diverse culture regions to the appropriate continent. A judicious selection of slides providing plenty of cultural clues serves to structure an activity reinforcing such a set of expectations.

The general notion is then challenged by a second round of slides showing the central areas of cities, from different parts of the world. It is much more difficult in this case to identify accurately the cities or their cultural regions since in commercial industrial aspects western style city architecture is spreading throughout the world (HSGP, Unit 3, 1969). The breaking down of the expectation forces a restructuring of the generalisation and an attempt at explanation. What features of city landscapes are remaining culturally distinctive? What features are bringing about uniformity? What contributed to shattering the expectations? One of the points which should be drawn out of this latter question is the selected and inadequate nature of the evidence upon which the conclusions were based. The discrepant data procedure helps students examine ideas held and gives them an opportunity to develop more precise and accurate generalisations; it helps to motivate them because they realise they held a wrong idea and they therefore need to make a fuller study; it helps them understand that they have certain values and that they have biases, stereotyped images and prejudices; and it helps them to become aware of the tentativeness of hypotheses, opinions, and generalisations and the need to check a variety of data sources before arriving at conclusions.

Discrepant data and the Third World

Professor Joachim Engel and his associates (Engel, 1980) constructed a discrepant data exercise to introduce a unit on Third World problems for the RCFP (The West German Geography/Social Science Curriculum Project).

In the first activity, after a word association discussion—'What do you think of when you hear the word, Africa, etc.?'—one readily identifiable slide of each of the major continents is projected. More than 90 per cent of a class match slide to continent correctly. A high confidence level of world knowledge is built up. The focus of attention shifts. 'Do you know what a developing country, a Third World country, is like?' In the ensuing discussion, the teacher listens, neither approving nor correcting and then shows eight slides out of which ones taken in Third World countries have to be identified. The number of correct selections is much lower. Why? The need for more precise information is appreciated and students realise that if they do not have this, their ideas and

perceptions of a country are likely to be inaccurate and undifferentiated.

Classroom activity

Broadening perceptions and aspirations

The lesson activities planned to broaden and fill out perceptions are built around a young Cameroonian boy, Tabi Egbe. Tabi Egbe reveals his lifestyle, his hopes and aspirations as a prelude to German students understanding the matches and mismatches between themselves and Tabi. In addition, the discrepancies between Tabi's aspirations and those of his school mates and the goals his society might set up are teased out. For Tabi Egbe, who lives in a small agricultural village, does not wish to be a farmer, yet city life may not offer him all he believes it will and it is in the country's best interest to encourage people to stay in the countryside.

In a research survey of reasons for school attendance and hopes for adult lives, 123 twelve-year-old Cameroonian boys and girls gave answers which fell into four main categories. In 18 per cent of the answers, obedience to parents' wishes to attend school and to work hard dominated. A similar percentage made statements along the lines that education opens up one's mind and prepares one for a worthwhile occupation—that of being an electrician, a nurse, or a teacher. Nearly one third indicated that they believed education would lead them out of a 'Third World life' into a 'First-World' one. A wealthy life in the city as businessman, state officer or president was a dream. Another third of the children went to school for the future benefits to be derived from being able to offer more to their society and being a better citizen. They saw these two goals as not only helping themselves, but their parents and country. In the Tabi Egbe unit, two suggestions for using this information are made. In one sequence, German students are presented with an overhead transparency cartoon of four children saying:

1. 'I go to school because I should like to help my parents.'
2. 'I go to school because I want to become a useful citizen to my country.'
3. 'I go to school because I want to become an important government official.'
4. 'I go to school because I don't want to be a peasant; I would rather like to get a job in an office.'

The German students are then asked to write down their reasons for attending school and typical answers compared with the ones from the Cameroon. A discussion follows of why students in a Third World country look similarly or differently upon their schooling and their futures compared with those in a rich industrial nation. The second suggested strategy is a reverse of the first. A German class is questioned about reasons for school attendance. When these have been collected and structured, the Cameroonian results are presented in a letter from Tabi Egbe. He writes to say that a German visitor has asked them to list reasons why boys and girls in the Cameroon go to school. In the letter he sets out the answers given by him and his friends. A comparison follows.

Value judgements

The final activity in the unit exploits a tension which is built up between the knowledge of 'Why they/we go to school' and the question, 'What should the school provide?' The German class is to be involved in planning for a better school in the Cameroon. A local Cameroonian teacher has succeeded in mobilising the community into building a new school close to the school farm. The teacher wants to run the kind of school which fits the needs as he sees them of young people in a rural area: the kind of school which might reverse Tabi Egbe's desire not to become a peasant.

In a letter to the education officer for the district, the teacher requests new equipment and materials. Of the thirty requests, listed in Figure 2.5, the education officer can afford to supply twenty and the teacher is asked to prune the list. Pruning the list is the task given to the German students. It is accompanied by the instruction:

> 'The new school should be so well equipped that parents will want to send their children. The children should learn things useful to their likely futures. Since living in the countryside would be better than living in the town, the school should prepare the children to be farmers with a knowledge of soil conservation, soil characteristics and crop management.'

German students must be prepared to defend the decisions they make. What value positions are they likely to take up? This is a crucial question, the implications of which are looked at in more depth in Chapter 4.

The consistent use of examining discrepant data or match/mismatch throughout the unit is clear and certainly highlights the real world, value laden

Figure 2.5 Making choices

From the list of 30 articles select 20 that the Cameroonian teacher should request from the government:

school benches	2 footballs
school tables	35 hoes
1 blackboard	seed-corn (maize)
10 books with stories about America, England and France	seed-corn (beans)
	spawns (coffee)
1 aquarium	fertiliser
5 stuffed birds	6 washing basins and towels
writing material for the pupils	1 ladder for gymnastics
1 wall-map of Cameroon	flower boxes for the window
1 wall-map of Europe	bowls and basins in which to keep harvest goods
2 microscopes	
1 set of scales with weights for experiments in agriculture	
35 saws	a shed for the tools and agricultural equipment
10 planes to work on rough wood	wooden posts for a fence
35 hammers	wire for a fence
nails, screws and hooks	5 garbage cans
	1 typewriter

dilemmas involved in evaluating and deciding upon best solutions. The validity of generalisations and policy decisions is open to forceful challenge and in this case there is a strong suggestion that it is the geography teacher's inescapable lot to raise rather than obscure the choice of the value positions taken on controversial issues. An example of a student's responses to a simplified version of the Tabi Egbe unit is given in Figure 2.6 and supports well the view that further value clarification and discussion is necessary and possible.

Professor Engel (personal communication) has pointed out that some of the more obvious distractors in the students' list include stuffed birds, the wall map of Europe, the aquarium and gymnastics ladder, flower boxes for the windows, garbage cans, the typewriter, microscopes and spawns. He recalls nevertheless, very good classroom discussions in which the apparently unnecessary items have been defended. A group of girls argued vehemently for flower boxes on the grounds that Cameroonian children had as much right to decorate their school buildings as German children. The list quite definitely

Figure 2.6 Education in a country with limited resources

There is a new school planned for a village in West Africa. There is not a lot of money available. Many of the children do not want to be peasant farmers like their parents. They want to be richer. The new school is to be equipped so parents will want to send their children. The children should learn things useful to their future in the countryside as *peasants* and *not* as *town dwellers*. BUT only enough money is available for 20 of the 30 articles the school board requested. Tick those items *you* would select:

school benches	2 footballs
school tables	35 hoes
blackboard and chalk	seed-corn
10 story books from America, England and France	bean seeds
	coffee bushes
aquarium	fertiliser
5 stuffed native birds	6 wash basins, towels and soap
paper and pencils	gymnasium equipment
wall-map of country and Africa	flowerboxes and seeds
wall-map of the world	storage bowls for harvested food
2 microscopes	tool shed
scales for weighing food	wooden posts for fences
35 saws	wire for fence
10 planes	5 garbage cans
35 hammers	1 typewriter
nails, screws and hooks	

Student answer

I have excluded (*10 story books from USA, UK, France, aquarium, 5 stuffed birds, 2 microscopes, 2 footballs, gym equipment, tool shed, wire for fence, 5 garbage cans and 1 typewriter*) because (*story books aren't necessary to being a peasant neither are aquarium and you don't need to know about birds to sow seeds. You don't need microscopes in the middle of the desert and footballs are a luxury for leisure and you can run round to keep fit instead of gym things. You can keep the tools in the school and you can cut down trees in Africa for fences and you can put rubbish straight into the disposal shoot. You can write in handwriting instead of print*).

Source: based on Ray Pask, RCFP.
has been successful in promoting discussion and encouraging students to submit convincing arguments and defences. Where teachers feel that one or more items may not prove good distractors in their classrooms then they should be replaced, e.g. a Japanese dictionary might be a better distractor in some cultures than a ladder for gymnastics.

Five faces of development

The British Schools Council Geography 14–18 Project based at Bristol also tackled the problem of providing teaching resources on planning and change in the Third World. The original Project Director, Dr Gladys Hickman, had a background of considerable field experience in Africa and selections from a unit on development is written up in the Project Handbook (Tolley and Reynolds, 1978). The aims embedded in the unit include developing an understanding of:

1. How social institutions and the *perceptions of decision makers* determine the allocation of resources to development.
2. How abstract ideas such as diminishing returns or models of economic growth *illuminate but never quite fit real situations*.
3. How values and cultures interact with other considerations in planning and implementing development priorities so that an *oversimplified technological view* of problems has to be avoided.

The specific learning activities are entitled, 'Five Faces of Development', which may be interpreted as illustrating five meanings, five levels, five contexts, five choices of development. Unexplored assumptions about what constitutes development is seen to provide a vast opportunity for mismatch. In the five activities, students move through:

1. an introductory map exercise which is a study of a developing area; to
2. studies in development as exemplified by small- and large-scale farming, and the role of an agricultural officer on the one hand, and Government on the other, in injecting change into farming; and finally to an exercise in deciding development priorities.

An interesting question sequence which could be adapted to suit any of the five faces is reproduced in Figure 2.7; likewise the simulation on development priorities in Figure 2.8. In the later case, the strategic similarity with the RCFP list pruning exercise is obvious and demonstrates that it is a very useful one for dealing with the concept of match/mismatch.

*Figure 2.7 The small African farm: how the socio-eco-
nomic system works*

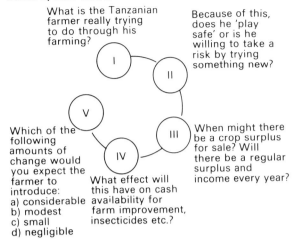

What is the Tanzanian
farmer really trying
to do through his
farming?

Because of this,
does he 'play
safe' or is he
willing to take a
risk by trying
something new?

I

II

V

III

IV

Which of the
following
amounts of
change would
you expect the
farmer to
introduce:
a) considerable
b) modest
c) small
d) negligible

What effect will
this have on cash
availability for
farm improvement,
insecticides etc.?

When might there
be a crop surplus
for sale? Will
there be a regular
surplus and
income every year?

SOURCE: Tolley, H. and Reynolds, J. B. (1978) *Geography
14–18: A Handbook for School-based Curriculum Development*, Schools Council/Macmillan.

*Figure 2.8 Development priorities and strategies— a
simulation for the Arusha area*

Introduction
1   Development plans are usually made by a team of
planning officers after much on-the-spot investigation.
The planning team's role is advisory. It should be the
task of representatives of the local people to decide
what parts of the plan should be implemented.
2   Although some of the development proposals listed
below are given locations on the Arusha 1:50 000 map,
all the proposals are simulated, not actual proposals.

Your task
   1 As geographical consultant your role is:
      a) to suggest the rejection of proposals which
         seem undesirable or impracticable;
      b) to recommend *three* projects for support.
   2 Each planning team will consist of 3–5 members.
      The team should reach agreement on what three
      proposals most merit support.
      Each team member should produce a report (no
      more than four pages in length) stating clearly the
      reasons for his rejection or recommendation of pro-
      posals.
   3 The money for the proposals must come from *local*
      taxes. Assume that a sum of £100 000 has been set
      aside for the three chosen proposals for each of the
      next five years, £500 000 in all.
   4 At the start of your report list five general criteria
      that you have applied in accepting or rejecting the
      proposals.

The development proposals
   1 No planned change—leave development entirely

to local individual initiative in adaptation to land
pressure.
2 Bring more land under cultivation by allowing
   farmers to clear woodland up to 7000-foot con-
   tour.
3 Consolidate all holdings of less than 5 acres into
   large units of 40 acres upon which maize or coffee
   could be grown using capital-intensive methods.
4 Move all farmers holding less than 5 acres to
   ujamaa settlements in less-populated areas of the
   Arusha region.
5 Build a dam across the Songota River at 494296,
   creating a large reservoir for irrigation.
6 Establish a government cattle ranch and meat-can-
   ning factory in the south-east corner of the map
   (i.e. the twelve grid squares bounded by 3525,
   3523, 4125, 4123).
7 Provide loans for local farmers to buy machinery
   and artificial fertilisers.
8 Provide loans for local people to establish small
   workshops and service industries in Arusha Town.
9 Build a large hotel (100 bedrooms) for foreign
   tourists on Sambasha Hill (square 4137) and improve the airfield (square 3527, 3617) to receive
   Charter Flight aircraft.
10 Take over the large estates and divide them up into
   20-acre units for progressive farmers whose
   present holdings are too small.
11 Extend the network of motor roads by widening
   and surfacing an additional 50 miles of tracks in
   the area north of the A23 main road.
12 Extend the agricultural advisory service by training
   more field officers.
13 Promote a family planning campaign.
14 Any other measures you think appropriate.

SOURCE: Tolley, H. and Reynolds, J. B. (1978) *Geography
14–18: A Handbook for School-based Curriculum Development*, Schools Council/Macmillan.

Priorities in development

The Christian Aid activity, 'Priorities in Development' (n.d.) should also be noted as an excellent
collection of data on the problem of choosing among
alternatives, though very often teachers simplify and
reduce the amount of information while retaining the
structure and purpose of the exercise. Christian
Aid's, 'Would you support these projects,' is another
shorter and telling exercise in deciding what requests
to meet in the face of an overwhelming number of
reasonable requests and needs from developing countries.

Mismatch as ignition

The Bristol Project has numerous examples of mismatch exercises scattered through its units to illustrate the clash of contrary perceptions, and the

implications for decision-making and reaching acceptable generalisations. Much data for such exercises, used as a starter or ignition resource or as the underpinnings of a unit of work, can be found once one has been alerted to the mismatch concept. Mismatch, a discrepancy of perceptions, values, information, expectations, often lies at the heart of geographical problems and therefore can be identified and used to integrate work and organise learning experiences and refine generalisations.

An issue of *Exploring Europe, Rural Depopulation* (1978), uses quotations and posters to highlight clearly different views held of depressed rural areas. Urban dwellers often feel that it would be marvellous to live in the countryside, while for the last 150 years and more rural depopulation has been persistent and widespread.

> ... despite our romantic urban notions the mountains are not the home of free and independent spirits but chiefly of depressed and poverty-stricken people, and the Good Life is not lived in desolate hills on National Assistance. All poverty is a prison, but poverty in the uplands is prison in a wilderness. The old way of life, whatever its antique virtues, is now rejected by those who know most about it, and improved local conditions have not persuaded hill people to stay.[1]

Some of the questions to be examined include:

Why the differing reactions?
Is there really a problem?
Will the problem go away?
How is the socio-economic system working?
What are the crucial dimensions of the problem?
How ought the problem to be tackled?
How can/should the problem be solved?

The clash of rural and urban perceptions has been presented as a fact which gives us the opportunity to search for and examine the general solutions available.

Different and incomplete information

To these strategies for challenging generalisations, another procedure may be added. It is so structured that generalisations based on partial and incomplete information have to be revised in the light of new evidence and Nicholas Helburn ascribes to it the acronym DAIS (Different and Incomplete Statements).

Classroom activity

Waste disposal and its environmental impact

How might students be introduced to such ideas and led to generalisations concerning perhaps the unity of the environment and the interrelationships between three forms of waste? A class may be divided into three groups (HSGP Unit 5 Habitat and Resources, Waste Management, 1969) or committees—one for each category of waste: solid, liquid and airborne. Specific roles are not necessary for students who can think of themselves as citizens with a duty to make recommendations on waste disposal. Each student is given a card with information about one of the three forms of waste and each is responsible for bringing the information and ideas to the attention of his or her committee. The cards feed in fairly specific ideas, on waste creation and disposal problems, the conflicts which can arise and some of the solutions which have been tried.

The committees meet separately to hammer out a report to be given to the class as a whole. Each report is to be structured around the following questions: What goes on now? What do you consider to be the worst problems? What changes do you recommend? What difficulties do you foresee in putting these changes into effect?

Modifying generalisations

Initially students have to summarise and come to some conclusions based on their individual assignment. Then those conclusions and general ideas are refined as, during discussion, ideas are seen to overlap, reinforce, develop, impinge upon other ideas. Understandings reached by individuals may be modified in the process of discussion and concepts and generalisations given a reorientation. It is an opportunity for expressive talking, talking to explore ideas. This process of learning through geography is elaborated in Chapter 5.

Finally, in class discussion of the three reports, two themes are basic and if these generalisations do *not* emerge the exercise has failed. First, the impingement of each kind of waste disposal on the others should come out. For example, if solid waste is burned, the dust could be a potential air pollution problem. Or if a dust catching mechanism is installed in the incinerator chimney, the dust may be worked into the sewer adding to the liquid waste problem. Second, the concept of the unity of the environment is important. The environmental impact of dumping waste on swampy areas is stressed in one or two of the individual assignments, for example.

This activity is carefully structured to build up the generalisations and decisions about one aspect of waste management. Those general ideas may then need to be changed as new perspectives are added from other reports and discussions. The strategy is simple and part of its appeal lies in its similarity to the way we come to change our ideas in everyday life. The strategy is also used in the Japan unit (HSGP, Unit 6, 1970) where again committees work with data to reach conclusions which are then built into wider generalisations, in the guise of recommendations.

Classroom activity

An exercise in route planning

One of the most ingenious exercises which has been circulating in geography's underground press for some time now is Rex Walford's 'Which route next? A problem for a Maritime Community of the future.' He has allowed its publication here. It is a striking example of a structured exercise deliberately building up from the simplest ideas and methods of analysis, to more complex analyses which allow the consideration

of the effects of more variables to reach a 'best', most realistic solution to a problem. Generalisations are modifed as the exercise proceeds.

Walford's scenario is set in the Maritime Provinces of Canada in the year 1997 and his verbal description and exposition is intended to humanise and give point to a numerical exercise which if presented as a bald arithmetic exercise could otherwise lose some of its impact and appeal. Undoubtedly, one of the problems attendant upon the quantitative revolution has been to persuade people as to its worth and relevance. The passage quoted below is from Walford's original exercise which has been changed only in small details. It is worth noting that it was devised for a group of geography students at McGill University. A guide to the necessary sets of calculations is set out in Figure 2.9.

Which route next?

It is the year 1997. Things have changed. . . . Since Quebec separatists finally gained their way in 1988, Canada has had to face the increasing challenge of 'Provinces Rights'. British Columbia and the Prairie Provinces (intent on holding the majority of their oil reserves against proposed Federal laws) followed Quebec in demands for greater autonomy. So did Newfoundland, now newly rich with the giant plutonium strike, made in 1987. Thus in 1994, the three Maritime Provinces found themselves to all

Figure 2.9 Which route next?

(a) What link should be added to achieve the fewest links in this network as a whole?

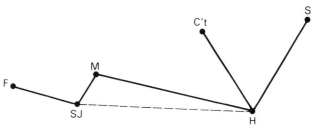

Calculations for a Saint John–Halifax link

Urban Centres	F	SJ	M	H	C't	S	Totals
F	0	1	2	2	3	3	11
SJ	1	0	1	1	2	2	7
M	2	1	0	1	2	2	8
H	3	1	1	0	1	1	7
C't	4	2	2	1	0	2	11
S	3	2	2	1	2	0	10
Total	13	7	8	6	10	10	54

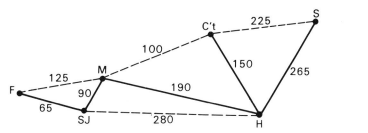

Calculations for a Saint John–Halifax link

Urban Centres	F	SJ	M	H	C't	S	Totals
F	0	65	155	345	495	610	1670
SJ	65	0	90	280	430	545	1410
M	155	90	0	190	340	455	1230
H	345	280	190	0	150	265	1230
C't	495	430	340	150	0	415	1830
S	610	545	455	265	415	0	2290
Total	1670	1410	1230	1230	1830	2290	9660

(b) What link should be added to achieve the least addition in mileage?

(c) What link is best in terms of likely traffic generation? The four possible links are:

Saint John–Halifax

$$\frac{310}{280^2} = 3.95$$

Moncton–Charlottetown

$$\frac{120}{100^2} = 12.0$$

Fredericton–Moncton

$$\frac{150}{125^2} = 9.6$$

Charlottetown–Sydney

$$\frac{150}{225^2} = 2.9$$

(d) What link is best in terms of all traffic flowing across the link?

A worked example:

$$M - C't = M - C't + SJ - C't + F - C't$$

$$12 \quad + \frac{120}{190^2} + \frac{70}{225^2}$$

$$12 \quad + \quad 3.3 \quad + \quad 1.4 = 16.7$$

intents and purposes the masters of their own fate, but unable to depend on revenues from elsewhere, since Federal taxes and grants were now abolished and Ottawa's power severely curtailed.

New Brunswick, Nova Scotia and Prince Edward Island have decided to go it together as a Maritime Community and they are seeking to integrate their economies, their public services and their governments. The new Prime Minister of this Maritime Community is a shrewd man; a geography graduate from Mount Allison University, he has been well-schooled in the ideas of Taaffe, Morrill and Gould. He knows that a nation is only as strong as its communications networks and that an integrated system is vital to the Community's future health.

Much travel these days is done by vertical take-off

and landing (VTOL) and one of his first priorities on taking office has been to set up an hourly service between the major cities of the Community. These cities are:

Fredericton	(50,000)
Saint John	(100,000)
Moncton	(100,000)
Charlottetown	(20,000)
Halifax	(210,000)
The Sydney region	(130,000)

But the Community (plus the new Maritime Airways) is chronically short of money. Thus the initial pattern of linkages from City centre to City centre is set up in the most economical way possible. The map of the initial network looks like this:

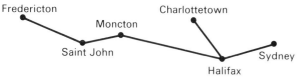

Now after two years, a chance to develop the network has come. There is enough money in the Maritime Exchequer to develop one further link in the chain, and the Maritime Parliament is discussing the matter. The M.P. for Glace Bay is first to his feet, pointing out the need for better communications for Cape Breton, and its need to be more closely integrated into P.E.I. (where the Parliament is temporarily meeting until a permanent home for the Legislature can be settled). He suggests the development of a Sydney–Charlottetown link....

The M.P. for Moncton North counters—'Clearly a more central route would be most helpful' he declares. Thinking of his own arduous route to the Legislature (via Halifax by VTOL)—still marginally quicker than by car and ferry—he proposes a Moncton–Charlottetown link....

The M.P. for Fredericton East then brings up a third proposal 'Let us cut down the extended western arm of the service by shortening a link' he says, proposing a Fredericton–Moncton link....

'Gentlemen', says the weary Premier (who comes from Halifax), 'Neither of these routes concerns my city—I clearly have no axe to grind—but I beg you, let us choose a route that benefits the Maritimers as a whole, not simply improves the advantage of any one city.'

(Here, he was being a little disingenuous, because all M.P.s knew that the more central the city on the network, the stronger the argument for it being eventually designated the capital.)

The arguing M.P.s then protested that they were not motivated by self-interest, and were concerned for the 'good of the network as a whole'.

At this point, the M.P. for Sackville leapt into the debate, 'Luckily I have been trained in the techniques of network analysis while a student at McGill University, and I think I can solve this problem', he said, passing a few scribbled calculations on an envelope to the Prime Minister whom he knew (as a fellow geographer) would understand them. 'I have analysed the network and its links, and undoubtedly the best link of the three proposed would be....'

1. What was the link that the McGill-trained M.P. suggested?
2. What were the tables of calculations that he showed the P.M.?
3. Was the P.M. (as an M.P. for Halifax) pleased on behalf of his own city, or not?

Some further episodes

An M.P. from Saint John (educated at the University of Toronto) was quick to challenge the apparently bewildering expertise of the Sackville member.

'Your calculations in links are the old crude approach to network analysis', he said. 'At Toronto they taught us to calculate in *linear* distances not simply in links. This gives a better, more refined set of calculations and incidentally changes the answer.' So saying, he handed an envelope with scribbled calculations on it to the P.M.

4. What were the second set of calculations?
5. How did they change the answer?
6. Was the Prime Minister (as an M.P. from Halifax) as pleased for his own city as he was before?

The M.P. for Sackville was not to be out-done—he leapt to his feet again. 'On the contrary', he declared, 'My friend is wrong. If we are going to calculate in something *other* than links, we should use not mere linear-distance, but the actual *time* that it takes to get between each place. This is not necessarily the same relationship as that of linear distance.'

7. What do you think of his argument as a general principle?
8. What do you think of his argument in this case?

Yet another M.P. entered the confused debate— this time a woman from Apple River, who had graduated at U.B.C. 'Out on the West Coast, we learnt about network analysis the proper way', she said. 'Surely the Honourable Members must realise that these calculations are invalid if we do not include calculations of the *amount* of traffic that will be *generated* by each link. Now the amount of traffic generated by each link depends on two things: (a) the size of the places at each end, and (b) the distance between the places, and we can set up a formula to show this. The simple application of this formula—known as the gravity model—can give us an indication of the potential flow of traffic on each link, and on *that* we should base our decision.

The formula is:

$$\text{Index of Flow between A and B} = \frac{\text{Population of A and Population of B}}{(\text{Distance between A and B})^2}$$

I have some calculations here.' And here a third envelope was handed to the P.M. who acknowledged the validity of the argument.

9. Did the third set of calculations reverse the decision again?

The M.P. for Little Shenogue now stood up. 'I hardly like to contradict my learned U.B.C. educated colleague', he said, 'But I learnt network analysis at the University of the Arctic in Yellowknife, and there at the newest of our Canadian universities naturally the methodology was the most up-to-date.

'Could I point out that the flaw in the last argument is that, for the calculations to be accurate, they must include *all* traffic flowing across the link—*not simply that generated by the two towns at the ends of it.* Thus the calculations for, say, Moncton–Charlottetown have to include not only the formula applied to those two towns, but also the formula of flow calculated for Fredericton–Charlottetown and St. John–Charlottetown, since traffic generated on those links would *also* use the new link.

I have an envelope here. . . .'

10. Did the fourth set of calculations reverse the decision again?

The Speaker of the House then tried to quieten the increasing disorder and hubbub. 'Please, please, *please*', he said. 'As a history graduate myself, I am quite lost in all this—what *are* we doing?'

'We are doing what any literate, numerate, civilised man would do', said the geographer P.M., scoring handsomely against his old historian rival. 'We are simulating a model of a network and refining it as we go—that is how we should seek sensible answers to our problems. . . .'

The undoubted relevance of this exercise to the general point about modifying and refining generalisations and decisions is clearly obvious and hopefully it will be adapted for use in other contexts and problems. It also provides a very convenient link into the third part of this chapter where the role of general ideas in decision-making and subsequent thinking is now stressed.

Generalisations for subsequent thinking

Decision-making, itself already part of earlier exercises, is highlighted once more by examples.

The first activity has direct appeal to many adolescents. It may be useful as an introductory lesson in some quite complex decision-making exercise. It derives from an environmental programme of some originality and foresight (Group for Environmental Education, 1973). The necessity to make a choice from alternatives each governed by constraints is clear. And the process of weighing-up the advantages and disadvantages of alternatives, parallels reaching generalisations, as relationships between needs and constraints, for example, not settlements and communications are explored.

The second activity fulfils a double purpose. It illustrates how general ideas contribute to decision-making and it is an example of a learning activity which shows decision-making to be based on both personal, subjective reactions and objective knowledge. It combines, to some extent, scientific geography and the understandings derived from such analysis with more humanistic responses. Above all, it illustrates the necessity for students to combine and weigh-up ideas.

Classroom activity

Locating a clubhouse

This first activity illustrating the nature of decision-making is directly addressed to students.

'Your club is looking for a clubhouse. You feel that you need a place where you can get together after school, during the weekends and the holidays. You have searched your neighbourhood for a place but so far every possible place has a disadvantage.

One place, the basement of a church, is ideally located in the centre of the neighbourhood but there are some real problems. You cannot use the room on Sunday morning, a time when you would be wanting to use it. The church wardens have also said that the room cannot be repainted and you were planning to be able to paint and decorate it the way you wanted to. Forty pounds for paint and brushes has already been set aside. The amount, together with a further forty pounds for the annual party, is your total annual income. Another place not too far from the church used to be a garage and workshop. It is now empty. It would be just right except that it would cost a lot to do up. As well as painting, new lights, a new door and possibly a new sink is needed. It could cost three times the amount of money saved. It would certainly mean giving up the party and there would be more money to find. There is a way to get the money but it would mean accepting new members. The club rules are against letting in any more members.

There is one more place, a vacant shop front, which you could use whenever you wanted and it could be decorated as you wish, probably at a cost of slightly less than forty pounds. The problem with this place though is that it is in another area and, as well as its being further away, there is not a great deal of friendship between you and the teenagers living there.'

Alternatives

What is to be done? No matter which place is chosen, you have to give up something—using it on Sunday mornings, decorating it the way you wish to, having the annual party or keeping the membership as it is.

Consider the following questions:

1. What do you think a clubhouse should be for?
2. What would you want to do there? List some of the different activities.
3. What kind of space would you need for these activities?
4. Are there any solutions other than a 'clubhouse' which could work?

Now let us look at the club's resources again. You have eighty pounds and members who are willing to work. Perhaps this is your most valuable resource. Can you think of any other resources—human or material—which were overlooked? Can you think of ways to increase these resources, human and material?

Who is to blame?

The members blame the church's rules for some of their problems. But there are other rules stopping the club finding other alternatives. What are they? Whose are they? Which rules would you change if you could and how would you change them?

When you first looked at the clubhouse problem you felt there was no way of getting a place to meet. You had three alternatives but each of them had a drawback. Now that you have considered the problem, you have possibly found new resources, perhaps changed some rules or thought of solutions not considered before. What choices does the club have? List them. Which would you pick? Why?

Personal decision-making

Turning to the second example of decision-making, this again emphasises reaching general understandings of a multi-variate problem. This example also illustrates further the complementary relationships between geography as science and geography as personal response.

Humanistic geography in its emphasis on the felt senses, the subjective knowledge of space is highlighting the bonds between people and places, landscapes and spaces experienced. It can be argued that it is through knowledge of one's specific, personalised environmental bonds, combined with knowledge of the objective reality of how things are arranged in the 'real' world, that one may sort out the best place to live according to income, life-style and personal preference. In teasing out the practical implications involved in buying a house, geography's objective and subjective dimensions are linked.

In the Metro housing game devised by Mhorag Ewing (1976), the outcome of the decision of where and what house to buy is assumed to contribute a great deal to a person's level of general satisfaction and so objective and subjective dimensions are fused in the game. The question sequence is a model of the decision-making process and the factors which need to be taken into consideration. It is based on a model by Brown and Moore (1968), identifying the order in which decisions are taken and the main factors ostensibly influencing the decisions. Their model postulates that at any given time an individual or household is liable to experience some form of stress. This stress is caused by any feature in the total environment bringing about discomfort, inconvenience or dissatisfaction.

The stress may emanate from within or without the household. For example, the decision to marry, the birth of a child, the decision to retire, the receipt of a large legacy, the conversion of a street to a one-way system, the arrival of undesirable neighbours, change in use of a nearby building, e.g. from office to off-licence, could all prompt a desire to live elsewhere.

The effect of the stress is dependent on the reaction to it, the strain felt. The personality characteristics of the household and its general housing and social experiences and aspirations and financial circum-stances will affect the outcome of the stress/strain balance. Most people do and, in fact, must, tolerate a certain amount of residential stress since few people are lucky enough to achieve total residential satisfaction. Strain may reach a threshold such that a decision is made to move or the strain converted into a decision to improve the existing environment by double glazing, joining a residents' protection association, etc.

If a decision is made to search for a new location,

Figure 2.10 Buying a house—an opening question sequence

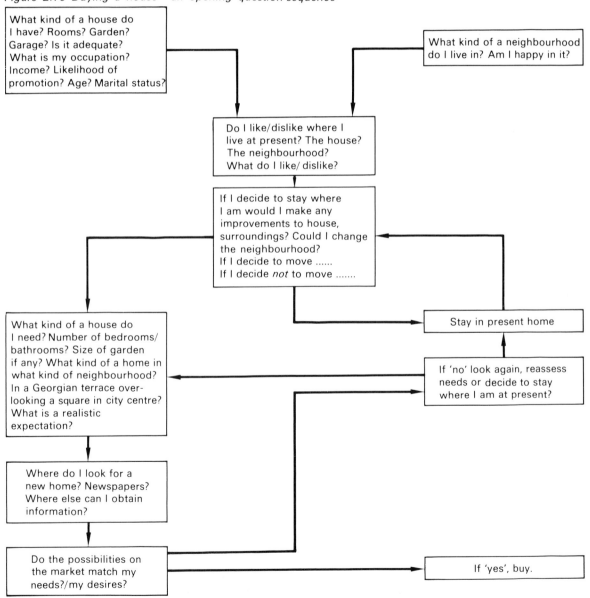

Source: based on Ewing, M. (1976) *Learning about housing and related issues: with specific reference to Metropolitan Montreal*, Unpublished M.Ed. monograph, McGill University.

then a series of plans and decisions outlined on the left of Figure 2.10 come into operation. The searching strategies (Wolpert, 1966) will concentrate generally among alternatives that are familiar, e.g. places nearby, areas where friends or relatives live, areas visited, areas described by the mass media or alternatives discovered by a systematic search.

Classroom activity

Metro housing game

The game 'Metro', devised by Mhorag Ewing as a role-playing game, focuses on the perceptions and preferences as well as on the sequence of decisions involved in house or flat purchase or rental. Each player in the game (with persona provided) becomes a family moving into a metropolitan area.

The game has two stages. In the first stage the family is dealing with and resolving the problem of searching for and buying a new house. In the second stage, during which the family must live in their chosen home for several years, they test their satisfaction with their decision against a number of events which can affect residents and residential areas. The unforeseen events are represented by chance cards involving family changes, governmental decisions, natural events, decisions emanating from neighbours and other groups of people in the housing system.

Sufficient information on the structure and the nature of the material is provided here to devise one version of several Mhorag Ewing suggested for 'Metro'. In this version all players are given the same role information and at the end of the game their decisions, house choices and satisfactions can be directly compared.

The game needs a real estate broker or agent who must:

1. explain and interpret the rules of the game;
2. show properties for sale;
3. audit mortgages;
4. control the market cards, offer the purchase slips, bidding slips and chance cards.

The teacher will probably decide to take on this role initially. The students are given a set of materials made up of:

1. family role information; (Figure 2.11)
2. maps of the physical land use and accessibility features of 'Metro'; (not provided)

3. The *Metro Home Buyer's Guide* giving step-by-step instructions for playing the game; (Figure 2.12)
4. *Metro News*, a real estate newspaper; (not provided)
5. chart for recording residential satisfaction and reaction to chance events. (Figure 2.13)

Figure 2.11 Metro Housing Game: a role description

Mr Bernard Goldberg and Family

You and your wife are in your early thirties and have a son and a daughter, both under twelve. Your new job will be as a salesman for Office Equipment Co., whose headquarters are in Hampton, close to the central business district. Your annual salary is $16,000 and you have saved $15,000 towards your new home. You have a grade 11 education but have also taken some business psychology courses in connection with your work. Your interests include dining out, coaching pee-wee hockey, going to football and baseball games and watching sports on television. Your wife's hobbies are dancing and gardening. Both you and your wife like to spend a lot of time and money on clothes and your appearance. For summer holidays you generally go to stay with your parents at their cottage, or sometimes your parents come to stay with you for one of the summer months. For work you drive a rented Pontiac while your wife drives a Mustang.

*   *   *

You have just arrived in Metro, a metropolitan area of approximately 500,000 people, located on the river Kelvin in the north-eastern area of North America. Metro has a humid, continental climate with long, cold winters and fairly heavy precipitation. Summers are hot and can be quite humid. The built-up area is expanding rapidly and there are several dormitory and industrial suburbs around the main city.

*   *   *

You are faced firstly with the problem of finding a home and secondly you have the objective of gaining satisfaction from living in that home over a period of years. Your sources of information concerning the city will be the *Metro Home Buyer's Guide*, *Metro News*, and several Metro maps. However you will also gain much information about the districts and housing types from your own experience of 'viewing' properties for sale.

SOURCE: Ewing, M. (1976) *Learning about housing and related issues: with specific reference to Metropolitan Montreal*, Unpublished M.Ed. monograph, McGill University.

Figure 2.12 Metro Home Buyer's Guide

Metro is a delightful area in which to live. It offers a wide variety of employment along with extensive recreational and shopping facilities. Properties listed in the real estate paper, *Metro News*, have been grouped by district to enable you to determine quickly the price ranges and types of housing available in any given area. Mortgage information is listed below.

Mortgage Information

When deciding what price of home you can afford, consult the following table. Consider your annual income and also your savings, which can be used for your down payment.

Amount to be borrowed ($)	Minimum annual income required to borrow this amount ($)
5,000	2,000
10,000	4,000
15,000	6,500
20,000	9,000
25,000	12,000
30,000	13,500
35,000	16,000
40,000	18,000
45,000	20,000
50,000	22,000
55,000	24,000
60,000	27,000

Procedure for Home Buying

Step One: Define your *family needs* and preferences by completing the Housing Requirements Chart below. Circle the appropriate choices.

Housing Requirements Chart

Family size:	2/ 3/ 4/ 5/ 6/ 7/ 8/ 9
Type of home preferred:	apartment/ bungalow/ 2-storey colonial/ Spanish style/ 2-storey traditional cottage/ duplex/ split-level/ other
Building material preferred:	brick/ stone/ stucco/ wood/ aluminium/ combination
Type of area:	as close to city as possible/ inner suburb/ outer suburb/ rural fringe
No. of bedrooms:	2/ 3/ 4/ 5/ 6
Heating:	oil/electricity/ does not matter
Age of home:	brand-new/ less than 5 years old/ 6–10 years old/ 11–20 years old/ age does not matter if in good condition
Garage:	none required/ 1/2

Step Two: Define your *budget* by considering your *income* and *savings* in relation to the mortgage information chart.
Maximum loan: $_____
Maximum house price you can afford: $_____

For example, if you earn $10,000 and have a savings of $10,000, you can use your $10,000 savings as your down payment and you can borrow $20,000 thus you can buy a home up to the value of $30,000.

Step Three: Study *Metro News* and the various Metro maps. From this information choose 6 homes that you wish to view. The exact location of each home is shown by the 6-figure grid reference system. *Plot the locations of homes that you wish to view* (plot locations on map 2).

Step Four: *View homes* with your real estate broker (*large index cards with photographs*). When you examine the information on the cards you can assess the nature of the homes and try to match the vacancies with your needs and budget. If you do not find a suitable home within one month (viewing of six homes represents one month) you must take a market card (red) which indicates fluctuations in price of real estate through time. You may then search for a second month, third month, etc. Each month that you do not find a home you must consult the market cards.

Step Five: When you see a home that you would like to buy within your price range you make a *bid* on it (fill in grid references, price, etc. on the yellow *offer-to-purchase cards*). Your broker will give you the *bid result* (*blue cards*) which tell you whether your bid has been accepted or not.

Step Six: When your bid is accepted, you move into your new home. Record which of the following statements is most applicable to your choice of home:
1. This is your ideal home at a price you can afford.
2. Close to your ideal house at a price you can afford.
3. Ideal home at a price slightly more than you can afford.
4. House basically satisfactory, although there are certain things you do not like; however price is right.
5. This home is not really what you want but you feel it is all that you can afford and perhaps you can move again later.
6. This is a nice home, but it is really too expensive for you.
7. You do not like this house but you feel you ought to buy quickly as prices are rising so quickly.

Step Seven: You must now *live in Metro* for a period of at least 8 years. During this time you will have the opportunity to discover some of the events both unforeseen and planned that happen within Metro. The events that happen each year are represented by *chance cards* (purple). It is up to you how many years you stay in Metro but you must stay at least 8 years. During this time you will record your changing satisfaction in the satisfaction chart below.

Step Eight: Consult your teacher for discussion and follow-up activities. (*Follow-up activities* are in Teacher Information Envelope.) (See also streetwork activities in Appendix I of Monograph 'Learning About Housing'.)

SOURCE: Ewing, M. (1976) *Learning about housing and related issues: with specific reference to Metropolitan Montreal*, Unpublished M.Ed. monograph, McGill University.

Figure 2.13 Residential satisfaction: perception chart

Improvement in satisfaction   Reduction in satisfaction

Chance Factor Number	Large	Moderate	Marginal	No Effect	Marginal	Moderate	Large	Decide to search for a new home
	☐	☐	☐	☐	☐	☐	☐	
	☐	☐	☐	☐	☐	☐	☐	
	☐	☐	☐	☐	☐	☐	☐	
	☐	☐	☐	☐	☐	☐	☐	
	☐	☐	☐	☐	☐	☐	☐	
	☐	☐	☐	☐	☐	☐	☐	
	☐	☐	☐	☐	☐	☐	☐	
	☐	☐	☐	☐	☐	☐	☐	
	☐	☐	☐	☐	☐	☐	☐	
	☐	☐	☐	☐	☐	☐	☐	
	☐	☐	☐	☐	☐	☐	☐	
	☐	☐	☐	☐	☐	☐	☐	
	☐	☐	☐	☐	☐	☐	☐	
	☐	☐	☐	☐	☐	☐	☐	

SOURCE: Ewing, M. (1976) *Learning about housing and related issues: with specific reference to Metropolitan Montreal*, Unpublished M.Ed. monograph, McGill University.

After reading the family role, described in Figure 2.11, the playing proceeds with the help of directions in the *Metro Home Buyer's Guide* as follows:

1. Consider the mortgage information.
2. Define family needs.
3. Calculate family budget.
4. Study *Metro News* and the features of the residential districts and plot locations of homes to be viewed on the city map.
5. View the homes with your real estate broker (homes are viewed by studying cards with photographs and other information about the resident. (Figure 2.14)
6. If you do not choose any of the three homes—it takes one month to view all three—then take a market card which will reflect variations in the market over time and search again. (see Figure 2.15)
7. When you have decided on a possible house make a bid for it on paper and hand it to the broker or agent, he will give you a bid result card which tells you whether your bid has been accepted or not. (see Figure 2.15)
8. If your bid is accepted you move into your new home and evaluate your choice against a series of statements in the *Metro Guide*.
9. You now live in 'Metro' for at least six years. Each year you receive a chance card (see Figure 2.13) and record your satisfaction on the Residential Satisfaction Chart.

Figure 2.14 Sample of possible homes to be viewed

HAMPTON

Well-kept condominium apartment building. Inside garage. Thermostat control in each apartment. 2 bedrooms.

LIVE RENT FREE
2 bedroom apartment $18,000

It is important to note that there is little direct competition with other players in this game. There are no defined criteria for 'winning'. The Satisfaction Chart is an original contribution to the art of game construction. The key debriefing questions include: On what basis did I make my choices? What factors affected my satisfaction or dissatisfaction?

SOUTHAM

3 bedrooms, needs some renovation, solid brick

SOUTHAM SPECIAL

$29,000

DURHAM

1960, 4 bedrooms, centre hall plan, beautiful garden. Air conditioning, de-humidifier installed recently.

A DELIGHTFUL FAMILY HOME – WELL PRICED

$42,000

KELVIN HEIGHTS

Detached stone, large treed lot, sunken
roman bathroom in master bedroom.
4 bedrooms. River view.

AUTHENTIC SIMPLICITY

$68,000

A ROMANTIC REFUGE

$90,000

IONA

6 bedrooms, very luxurious residence
fronting on the Kelvin River,
large landscaped lot with mature trees.

MOUNT FOREST

5 or 6 bedrooms. Detached stone,
beautiful view large reception hall,
living room, dining room,
modern kitchen, upstairs library,
2 open fireplaces.

PRIVACY WITH ROOM FOR ALL

$150,000

SOURCE: Ewing, M. (1976) *Learning about housing and related issues: with specific reference to Metropolitan Montreal*, Unpublished M.Ed. monograph, McGill University.

Figure 2.15 Samples of market cards, bid result cards and chance cards

Prices are up $2,000 on all properties

Prices remain stable as advertised.

Taxes have risen sharply in your municipality. You do not feel the extra cost is justified. Lose satisfaction.

The government has 'frozen' all development on agricultural land in an effort to prevent further urban sprawl. If your home is already built in an agricultural area (see map 1) you can be sure that the area will maintain its pleasant character. Gain satisfaction.

The land opposite your home has been acquired by the city to make a park.

The owner will only accept the asking price and nothing less.

If you can state (to the satisfaction of the broker) two reasons to justify why the home is not worth the asking price, your bid will be accepted.

The city proposes to widen your street. As a result, you will lose 3 m from the front of your property. Lose satisfaction for loss of privacy and quietness, due to the closer sidewalk and street.

Economic Crisis
Mortgage money becomes very scarce. All the 'Amount to be borrowed' totals shown in the Mortgage Information Chart are reduced by one-third.
   If you could formerly find a mortgage of $15,000, this now becomes $10,000.

Beamsville, Greenwich, Hampton only
The church at the end of your street has been converted into a bingo hall. This increases the traffic on your street and lowers the amenity value.

Sorry.
Owing to illness, the owner has temporarily withdrawn this property from the market.

New neighbours have recently moved into the house next door. You find that their attitudes and interests are very similar to your own.

SOURCE: Ewing, M. (1976) *Learning about housing and related issues: with specific reference to Metropolitan Montreal,* Unpublished M.Ed. monograph, McGill University.

Debriefing games

Debriefing is an essential and integral part of choosing to use a game or simulation. Failure to debrief may leave students with distorted views. The following questions provide a guide to debriefing:

1. What happened in the game? What were the goals? What strategies were effective in accomplishing those goals? Which strategies had negative effects? It is sometimes useful for students to keep diaries of what went on.

2. What would happen if the rules or values were changed/or the penalties or rewards changed? How would this have affected your actions?
3. How did the game or simulation compare with reality? What additional factors would have made the game more realistic? How could the game be redesigned to make it more realistic?
4. Did the outcomes of the game seem fair? Was this the fault of the game or reality?
5. What hypotheses about reality did the game suggest? What would need to be done to confirm these hypotheses?

6. Did the game go against any of your values?
7. Should the game be followed up by other materials, readings or films?

Evaluating games

An evaluation of the game itself may be undertaken by applying Gordon Elliott's criteria (1975) reproduced in Figure 2.16. It is a clear and simple chart covering the main questions which are likely to arise in selecting and evaluating games. It should be added in passing that games, simulations and role plays are often organised on a group basis and they are one of the few teaching strategies which are genuinely suited to fostering co-operative learning in mixed-ability classes (provided one entertains and accepts the assumptions of mixed-ability grouping) (Slater and Gallagher, 1979). A lengthy discussion of grouping

strategies and class organisation lies beyond the reach of this book on planning learning activities but the diagram in Figure 2.17 on the problem-solving techniques in a group situation from *Motorway* (Rawling, 1976) is an example of planning for learning through geography in groups. Spicer (1976) provides another model which is discussed in Chapter 3.

Figure 2.16 Evaluating a game

Questions	Parts of a game	Criteria	Score
1 What is the central problem presented in the game?	problem	clarity conceptual content utility relationship to real-world	
2 What choices are available to players?	choices	soundness	
3 What are the different moves or activities provided for players?	moves	consistency	
4 What are the rules for the game?	rules	lack of distortion	
5 How is the game organised?	organisation	inclusive, sequencing, relationship to choices, moves, rules	
6 What summary activities conclude the game?	conclusion	adequacy, applicability, relationship to activities	

SOURCE: Elliott, G. (1975) 'Evaluating classroom games and simulations', *Classroom Geographer*, October.

Figure 2.17 The problem-solving technique in a group situation

Stages in problem solving	Group work
Choice of problem.	Which is the best route for a new Motorway to link London with the Severn Bridge?
Identification of problem by defining it.	Formation of class into groups of five. Study of background information about the area under consideration. Allocation of roles and examination of memoranda.
Collecting, organising and evaluating relevant information.	Individual study of the problem from a single specialised aspect by each separate member of the group. Preparation of two alternative routes.
Formulating hypotheses.	Board meeting. Group meets and the co-ordination of diverging viewpoints is attempted. Group's final proposal is decided.
Critically analysing by comparison and discussion.	Presentation of the group's proposal to the class and the Department of the Environment. Comparison with rival proposals and discussion.
Evaluation of solution and acceptance or rejection.	Final evaluation and decision by the Department of the Environment, and comment from the class. Acceptance or rejection of each proposal within each group.

SOURCE: Rawling, E. (1976) *Motorway*, Geographical Association.

Conclusion

This chapter has chiefly been concerned to stress the need to plan learning activities towards general understandings. A recent British publication (*Teaching ideas in geography*, 1978) has detailed many of the possible generalisations geographers may work towards. Readers may like to refer to the booklet. The idea of generalisation-making is a recent development in classroom geography in Britain and interested readers should go back to the American HSGP materials for early formulations of generalisations.

The idea has been developed in this chapter that working towards generalisations may be seen as a three-stage process, sometimes with all stages being present in an activity, sometimes with one or other being dominant. Examples of learning activities have been described to illustrate the feasibility of such a view being put into practice. Planning to teach towards generalisations, to modify or restructure generalisations, and to use generalisations in subsequent thinking produces learning activities in which meaning is developed and cognitive and social abilities exercised and horizons broadened.

In addition to the emphasis on generalisations and decision-making, the strategies emphasised throughout the chapter can also be held up as particular pieces of evidence that geography teaching provides students with a chance to develop thinking skills serving both intrinsic and instrumental educational goals. It is to the development of intellectual, social and practical skills that we shall turn in the next chapter. The identification of appropriate questions and answers represents the first and penultimate steps in activity planning. What resources have we from which we identify questions and reach generalisations? What data must be presented to students and what skills need to be developed to promote learning through geography?

Further reading

There are few articles or books which *stress* reaching generalisations and decisions as their *main* and explicit theme.
1. Any of the units of HSGP may be successfully analysed from the above viewpoint as the following

analysis of Metfab indicates.

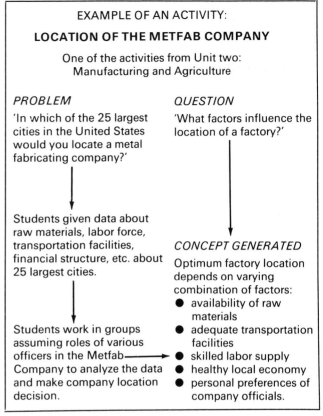

EXAMPLE OF AN ACTIVITY:

LOCATION OF THE METFAB COMPANY

One of the activities from Unit two:
Manufacturing and Agriculture

PROBLEM
'In which of the 25 largest cities in the United States would you locate a metal fabricating company?'

QUESTION
'What factors influence the location of a factory?'

Students given data about raw materials, labor force, transportation facilities, financial structure, etc. about 25 largest cities.

CONCEPT GENERATED
Optimum factory location depends on varying combination of factors:
● availability of raw materials
● adequate transportation facilities
● skilled labor supply
● healthy local economy
● personal preferences of company officials.

Students work in groups assuming roles of various officers in the Metfab Company to analyze the data and make company location decision.

SOURCE: Gunn, A. M. (ed.) (1972) *High School Geography Project, Legacy for the Seventies, Centre Educatif et Culturel*, Montreal.

High School Geography Project of the Association of American Geographers. (1965, 1979)
Geography in an Urban Age (Macmillan)
Unit 1—Geography of Cities.
Unit 2—Manufacturing and Agriculture.
Unit 3—Cultural Geography.
Unit 4—Political Geography.
Unit 5—Habitat and Resources (1st Edition). Environment and Resources (2nd Edition).
Unit 6—Japan (1st Edition only).
See also Gunn, A. M. (ed.), *High School Geography Project, Legacy for the Seventies* (Centre Educatif et Culturel, Montreal, 1972).
Gunn, A. M., (1975) 'The role of the High School Geography Project in mass education' in *Journal of Geography*, Vol. 74, No. 5, pp. 265–74.

2. It is profitable to examine project materials and chapters of textbooks to test the extent to which they

fit a generalisation-generating model or can be re-arranged so as to fit such a model. Examine the British Schools Council 14–18 materials and the British GYSL materials in this way.

3. Examine the lessons and activities to be found in any of the following (books or pamphlets only referenced, see Chapter 3 for journals):

Bacon, P. (ed.), (1972) *Focus on Geography*, 40th Yearbook, National Council for the Social Studies.

Ball, J., Steinbrink, J. and Stoltman, J. (eds.), (1971) *The Social Sciences and Geographic Education*, Wiley.

Geofile (1976) Charden Publications, Queensland, Australia.

Graves, N. J. (1980), *Geographical Education in Secondary Schools*, Geographical Association.

Graves, N. J. (ed.), (1982) *UNESCO Handbook for Teachers of Geography*, Longman.

Manson, G. A. and Ridd, M. K. (1977) *New Perspectives on Geographic Education: Putting Theory into Practice*, Kendall/Hunt.

Marsden, W. E. (1976) *Evaluating the Geography Curriculum*, Chapter 15, 'Units', Oliver and Boyd.

Resource Units for Geography Teaching, Geography Resource Centre Teachers' College, Christchurch, New Zealand.

Secondary Geographical Education Project (1977) *SGEP-PAK*, Geography Teachers Association of Victoria.

Teaching Geography (occasional paper series), Geographical Association, England.

Tolley, H. and Reynolds, J. B. (1978) *Geography 14–18: A Handbook for School-based Curriculum Development*, Macmillan.

Walford, R. (ed.), (1973) *New Directions in Geography Teaching*, Chapters 1–12, Longman.

Walford, R. (ed.), (1981) *Signposts for Geography Teaching*, Section 2, Longman.

4. Finally, for general reading on some of the issues raised in this chapter, refer to:

Bennetts, T. (1972) 'Objectives for the teacher', in Graves, N. J. (ed.), *New Movements in the Study and Teaching of Geography*, Temple Smith.

—(1973) 'The nature of geographical objectives', in Walford, R. (ed.), *New Directions in Geography Teaching*, Longman.

D.E.S. (1978) *The Teaching of Ideas in Geography*, H.M.S.O.

Dinkele, G. (1978) 'Reformed geography: a statement to whom it may concern', in *Teaching Geography*, Vol. 1, No. 1, pp. 13–14.

Fitzgerald, B. P. (1968) 'The American High School Geography Project of the Association of American Geographers', in *Geography*, Vol. 54, No. 1, pp. 56–63.

Geipel, R. (1979) 'Curriculum development and society: West German geographers respond to the American high school geography project', in *Journal of Geography in Higher Education*, Vol. 3, No. 1, pp. 84–5.

Hall, D. (1976) *Geography and the Geography Teacher*, Allen and Unwin.

Helburn, N. (1968) 'The educational objectives of high school geography', in *Journal of Geography*, Vol. 67, No. 5, pp. 274–81.

Jakubek, O. (1971) 'Geographic investigation of a social problem: a study in scientific method', in Ball, J., Steinbrink, J. and Stoltman, J. (eds.), *The Social Sciences and Geographic Education*, Jacaranda Wiley.

Kaltsounis, T. (1970) 'An analysis of teaching strategies in emerging geography', in Bacon, P. (ed.), *Focus on Geography*, 40th Yearbook, National Council for the Social Studies.

Kurfman, D. (ed.), (1978) *Developing Decision-making skills*, National Council for the Social Studies.

Marsden, W. E. (1976) 'Principles, concepts and exemplars, and the structuring of curriculum units in geography', in *Geographical Education*, Vol. 2, No. 4, pp. 421–9.

—(1979) 'The German Geography Curriculum Project', in *Teaching Geography*, Vol. 4, No. 3, pp. 128–30.

Natoli, S. J. and Ritter, F. A. (1979) 'The revision of the high school geography project', in *Journal of Geography in Higher Education*, Vol. 3, No. 2, pp. 102–5.

Slater, F. A. and Spicer, B. J. (1977) 'Objectives in sixth-form geography: towards a consensus', in *Classroom Geographer*, December, pp. 3–10.

Walford, R. (1973) 'Decision making', in Bale, J., Graves, N. J. and Walford, R. (eds.), *Perspectives in Geographical Education*, Oliver and Boyd, pp. 211–21.

3 Reaching Generalisations and Decisions through Processing and Interpreting Data

Codifying understanding

In the last chapter, a spotlight was directed at the importance of working towards generalisations and decisions. The generalisations and decisions developed in each exercise or activity provided answers to questions in an enquiry sequence. Reaching generalisations and making decisions is equal in importance to listing a set of questions in order to reach take-off point in lesson planning.

Generalisations complete an enquiry by codifying understanding. The threads of our understanding are drawn together, inter-related and stored within a generalisation or made manifest in a decision. In lesson planning, it is as important to work towards a general understanding from a question as it is helpful to initiate an enquiry sequence through a question. The model of lesson planning being outlined is starkly simple:

Identify question – – – – ► Develop generalisations

Such a progression is crucial as a structure for activity planning and as a strategy for developing meaning and understanding. Meaning and understanding define the process of tying little factual knots of information into bigger general knots so that geography begins to make sense, not as a heap of isolated facts but as a network of *ideas and procedures*.

The purpose of generalising

Emphasis was also given in Chapter 2, to two major considerations within the task of generalisation making: (1) refining generalisations, and (2) using generalisations in subsequent thinking. In this chapter, the emphasis is placed on a range of data forms as the evidence on which the generalisations are based and from which they are teased out.

How are generalisations formulated? What are some of the forms of data and data processing methods used to learn through geography? Examples illustrating some of the many possible answers to these questions provide the main body of the present chapter.

The nature of data processing

Searching for an answer to the question, 'Why won't my car go?' necessitates a specific examination of each spark plug (Collingwood, 1939). For Collingwood, there is a considerable amount of information to be examined, searched through, processed and interpreted, in order to move from question to answer. The spark plugs are viewed as analogous to maps, statistics or whatever form of data we are working on.

In fact, as geography teachers we have a rich bank of data or resource items to aid us in the process of helping students achieve generalisations. Figure 3.1, expanded from documents published by the New Zealand geography curriculum renewal programme, lists some of the data forms available in a bank as it were. It suggests the range of our borrowing capacity and along with this the opportunities that exist for arousing interest and motivation through the selection and presentation of data in a variety of forms. After presentation, Collingwood's analogy implies, the data must then be processed.

In examining the spark plugs, Collingwood had to select a number of strategies from among available data gathering and processing methods. Undoubtedly, he somewhat unconsciously selected observation as one method and proceeded to collect and mentally record what he observed. After observation, he had then to analyse and evaluate the data by sorting out and recognising significant elements in

Figure 3.1 Bank of possible data/resource materials

Diagrams	Photographs	Models
Maps	Posters	The Field
Newspaper articles	Cartoons	People
Advertisements	Sketches	Cassette recordings
Periodicals	Case Studies	Films/Filmstrips
Documents	Statistics	Videos
Texts	Graphs	Computer Programs
Resource packs		

SOURCE: based on an idea from the New Zealand geography curriculum renewal programme.

his evidence on the performance of the spark plugs. The evaluation may have enabled him to make a general statement about the relationship of individual spark plugs and engine performance.

Skills in data processing

The Collingwood example has been elaborated on in order to stress the general idea that data processing methods have to be applied to generalisation-making and decision-forming tasks rather than generalisations being handed out to be learned by rote. It follows that in data processing, the learner applies a number of skills or goes through a set of tasks to reach a conclusion. The idea of the simultaneous impingement of methods and skills on data to form generalisations may be illustrated thus:

Intellectual, social and practical skills— overlapping categories

Skills are exercised in applying methods to tasks involving data gathering and processing. Many teachers have their own checklists of the skills and tasks most often used in geography teaching. Figure 3.2, again inspired by the New Zealand programme, is probably broadly representative of such lists. A division into intellectual, practical and social skills has been widely adopted in recent years, often in the context of discussions about objectives.

The threefold classification is not as tight as some might wish and raises questions about where, for example, skills of communication are best placed. Communicating through talking or writing, for example, requires considerable intellectual ability but the arts of communication are also social skills or skills which serve social ends. At a fairly low level, talking and writing are also practical skills in that one cannot deny the elements of physical co-ordination involved in the production of both. It seems a decision on where to place communication skills can be made on the grounds that the bulk of talking and writing in a formal educational setting is a social skill serving intellectual and social purposes. Hence, communication skills can be classified as social skills though in Venn diagram style there is considerable overlap with the other two categories.

Study skills such as using an atlas, finding resources and organising time are also sometimes separated out into a special category. Again, the overlap between categories is great and seems greatest with practical skills. Finding resources is an exercise in practical know-how. Co-ordination of knowledge is involved, as well as the co-ordination of eyes and fingers in sorting through a catalogue, for example. But on balance, the exercise of skills when a secondary student has mastered the task of looking up a catalogue or searching through an index seems more practical on the whole than intellectual, though this argument is by no means watertight. Intellectual skills, I consider, predominate in the *use* made of the resources and references. The overlap arises in that practical skills are acquired and used in social settings for intellectual purposes towards educational ends. However, the chief significance of such classifications is that they are reminders of the variety of tasks

Figure 3.2 A bank of possible data processing skills/ strategies/tasks

Intellectual skills

1. perceiving and observing
2. memorising and recalling
3. understanding instructions/information
4. structuring information, classifying and organising
5. questioning and hypothesising
6. applying information and ideas
7. elaborating and interpreting
8. analysing and evaluating
9. identifying and synthesising
10. thinking logically, divergently, imaginatively
11. thinking critically and reflectively
12. generalising, problem solving and decision making
13. clarifying and analysing attitudes and values
14. communicating facts, ideas, concepts, arguments, results, values, decisions, feelings.

Social skills

15. communicating and planning *with others*
16. participating in group discussions
17. listening to other viewpoints and opinions
18. adopting a role
19. exercising empathy
20. working independently
21. helping others
22. leading a group
23. participating in field or research work
24. exercising choice and discrimination
25. behaving responsibly and courteously
26. accepting responsibility for learning
27. initiating and organising a learning task

Practical skills

28. talking, reading, writing, drawing, acting . . .
29. manipulating instruments and equipment
30. finding books and resources
31. walking an urban trail
32. using a map
33. organising a field investigation
34. preparing a wall display
35. administering a questionnaire
36. interviewing a town planner
37. designing a graph
38. taking photographs
39. sketching a building
40. presenting statistics
42. writing a report.

SOURCE: based on an idea from the New Zealand geography curriculum renewal programme.

which are all part of what learning through geography means. They are tasks which develop intellectual, social and practical abilities, so contributing to successful learning, conceptual development, meaning-making, increased understanding, awareness of and eventual participation in society's decision-making processes.

A very useful and practical extension of the concept of banks of skills and data forms has been reported by Macaulay (1979). The checklists of skills and data can be cross-referenced to provide yet a third type of checklist in the form of a matrix. This third checklist/matrix illustrates the possibilities for translating statements or descriptions, for example, into sketches, diagrams or maps. Similarly, sets of figures may be translated into statements, tables or graphs. This idea for the translation of data from one form to another is illustrated in Figure 3.3 and could provide a useful monitor to the mix and balance of the presentation of data being planned in activities. Any teacher wants to avoid an unvaried diet of data forms and tasks.

Resources for learning

The emphasis given here to the variety of (1) data forms, (2) methods of processing data, and (3) skills to be applied to data may serve as a reminder of the resources available to us to promote generalising and decision-making in learning through geography. But what does this add to planning considerations? There are now three distinct parts to planning learning activities:

1. identifying questions; (Chapter 1)
2. reaching generalisations; (Chapter 2)
3. processing and interpreting data. (Chapters 3 and 4)

The order is more appropriately rearranged to read:

Identify questions – – – –► Process data – – – –► Develop generalisations

Selecting student activities

In the previous chapter a modified version of the SGEP model of planning set out a sequence of tasks to be undertaken in arranging learning activities. The instruction to define objectives in box 9 in Figure 2.2 has been discussed in the light of the particular learning tasks and broad generalisations mirroring specific

Figure 3.3 A translation matrix

	Statements	Photographs	Sketches	Maps	Statistics	Diagrams
Statements, descriptions, accounts – – –						
Photographs, slides – – –						
Sketches, cartoons – – –						
Maps – – –						
Statistics – – –						
Diagrams, models, flow charts – – –						

and general objectives. Now, Figure 3.4 presents a six-step plan for enquiry-based lessons and identifies the next steps to be tackled in planning as selecting resources and devising student activities and teaching

Figure 3.4 Key steps in planning enquiry based lessons

STEP ONE:	Identify questions.
STEP TWO:	Decide what answers, what generalisations to work towards.
STEP THREE:	Gather and select appropriate data and resources.
STEP FOUR:	Sort out how to present the content and use the data: What learning tasks? What teaching strategies? Is there a balance and range of tasks?
STEP FIVE:	Examine the activity for likely educational objectives. Accept, reject, modify the activity.
STEP SIX:	Devise assessment and evaluation procedures.

strategies. An exercise designed by some geography teachers in Australia, using the question identification approach to planning, illustrates a judicious choice and amalgam of data forms, methods of processing and application of skills. The student activities, slightly modified from 'How well placed is your school?' are tabulated and briefly described below (Blachford *et al*, 1976).

Classroom activity
How well placed is your school?

As an initiating activity, students presented with two hypothetical dot maps as in Figure 3.5 are asked:

1. How well placed is the school from the point of view of students?
2. How well placed is the school from the point of view of teachers?
3. What would be the best place for the school?
4. What difference would it make if some students had to walk and others were driven in cars?

Figure 3.5 How well placed is your school?

Suppose a school had 30 students, as shown by dots in the sketch below

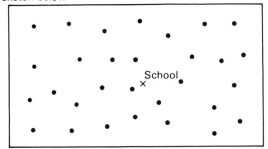

What if the 30 students lived in the area in the following way?

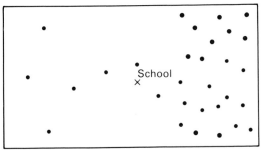

In this case what would be the best place for the school?

SOURCE: Blachford, K. *et al* (1976) *Way Out*, Rigby.

Maps form the data base for the student tasks of observation and interpretation.

The second activity requires a survey to be made of the class as well as other classes and teachers in the school. Three questions are to be asked:

1. How far do you live from school?
2. In what direction do you live from school?
3. How much time does it take to get to school?

The results of the survey are plotted as illustrated in Figure 3.6 on

1. a distance diagram, and
2. a time diagram.

Examining data

Data are collected by a survey and students are required to plot dot distributions in order to examine patterns. The examination is directed by questions. General conclusions about 'How well placed is your school?' can be reached by completing the following tasks or questions with reference to the data:

1. Is there an even spread of people all around the school?
2. Are there any parts of the diagram which have clusters?
3. Are there more people closer to the school or are there more people further away?
4. Are the students in your class spread about in a different way from other students and teachers?
5. How many people do you estimate are close to school?
6. Are there many people who are not far from the school but who spend a lot of time getting there? For what reasons?
7. Are there people who do not spend much time getting to school but who are a long distance from the school? Why?
8. What is the usual time taken and distance travelled to school?

Figure 3.6 Distance and time diagrams

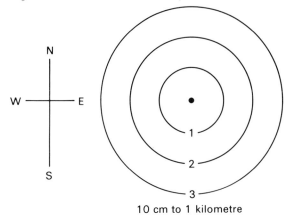

10 cm to 1 kilometre

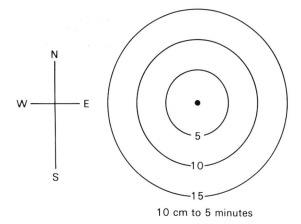

10 cm to 5 minutes

SOURCE: Blachford, K. *et al* (1976) *Way Out*, Rigby.

Analysing patterns

Students are being asked to analyse patterns and use such concepts as time, distance, proximity, distribution, pattern, clustering, accessibility and average, to make a general judgement about the location of the school. Data have been collected and processed by using a number of intellectual, social and practical skills to reach a conclusion in answer to a question. The work is arranged in a logical progression from an introduction to a development and conclusion with learning activities appropriately chosen to elaborate and reinforce points. The linking and intermeshing of data and skills to produce a generalisation illustrates the balance which should exist between the three elements, (1) question, (2) data processing, (3) conclusion/generalisation.

Subjective and objective data

Data collected by a survey yield what might be classified as objective data. Certainly in the exercise above it has been treated objectively. Subjective data can also be used, however, in learning activities and indeed it is such personal data and information about our environment and environmental experience that humanistic geographers claim we are in danger of ignoring. This point was brought home to me most strongly during a post field trip exercise. The exercise was a perception type of test modelled on one Hamelin had devised as a research procedure on perceptions of the Canadian North (Hamelin, 1972; Slater, 1976).

Classroom activity

Field work on the West Coast

A party of fifteen-to-sixteen-year-old girls living in the Oamaru district of the East Coast of the South Island of New Zealand had spent three days of a week-long field trip on the West Coast, the character of which is beautifully described in the following passage:

> The Coast, as it called itself, was a place for rain, beer and coal, almost to extravagance. There had also been gold in the fast rivers which rode from the white mountains. But this gold now lay locked in distant bank vaults and blighted towns were vanishing back into the bush. There was also, of course, a still earlier time, when only dragons of legend and thin, defeated tribes inhabited this slender length of lowland. Stronger tribes came from the north, to gather the greenstone which lay thick in the rivers and to harvest slaves. The Coast had never been in on a win. Now they took the coal away. There were still the mountains, though, in dense dozens above the bush, sometimes pushing peaks almost to the sea, and throwing off glaciers. There were the mountains, and the icy winds off them, and the heavy rains they trapped, and there was the creepered rain forest and the crashing sea; and there were the places where men felled timber, or dug for coal, because there was not much else.[1]

Resource exploitation

I had drawn up a list of words typical of the Coast which were functionally related on a resource utilisation basis—one of the dominant themes in the geography of the West Coast and one then frequently examined in the New Zealand School Certificate examination generally taken at fifteen plus. The words were as follows: 'sluices', 'Maoris', 'gold', 'Europeans', 'coal', 'aerial cableways', 'rimu', 'tourists', 'glaciers', 'bush', and 'greenstone'. My intention was to have them group the words by resource association, e.g. Maoris and greenstone.

Discrepant perceptions

It became apparent that the twenty girls who had turned up voluntarily in the lunch hour for the debriefing exercise were not happy with the list. To them, the words were not at all representative of the dominant features of the geography of the Coast. It then seemed more to the point to have them suggest words for a list and as a result of an informal, free-for-all discussion, for which I acted as recorder, the following list of the most popular words and phrases was compiled: 'glaciers', 'dredge tailings', 'rain', 'dirty rivers', 'rough pastures', 'bogs', 'lack of roads', 'lack of farms', 'hills', and 'bush'.

They were most conscious of either (1) what the Coast lacked, *vis-à-vis* their own East Coast which was prosperously farmed, though drought prone, and had plenty of well-graded roads and high quality pastures, or (2) aspects of the natural environment which were most distinctively different, such as glaciers, bush and the frequency of rainfall. By modifying the exercise as the girls wished, they and the teacher learned a lot more about their perceptions, subjective impressions, and levels of thinking.

Their experiences differed considerably from my geographically trained adult view. The exercise became an opportunity for revealing their private geographies of the Coast at that particular stage in their personal and intellectual development.

It was more disconcerting to find that in response to 'How do you see the West Coaster?' the group produced the following list: 'conscientious', 'sensitive to his or her environment', 'boring', 'impolite', 'ungracious', 'parochial', 'aggressive', and 'has strong regional attitude'. Some of these answers are indeed unintended learning outcomes and unplanned results of the formal pre-arranged meetings with national park rangers, farmers, and local councillors and the informal contacts with shop owners.

I have no doubt, despite their initial impressions and private geographies, that the girls could have written excellent answers to an examination question on the resources of the West Coast. They were capable of acquiring and using the right sort of framework and information for examination purposes.

I am convinced, however, that much of the lack of interest and motivation teachers experience is related to our not being sufficiently aware of and sensitive to 'where the students are at' before we begin the work of linking their experiences to other ways of viewing and structuring knowledge. I can recall a recent complaint from PGCE students about the disruptive and even hostile behaviour they observed in a group of inner city English girls during a field week in a national park. Each day was devoted to a hypothesis testing exercise. Few of the girls had ever been beyond London let alone in the strange environment of a semi-wilderness area. What might have been their attitudes and interests if some attempt had been made to begin at their beginning and simply orient them to the area through short pleasant walks with an emphasis on sensory experience and recreational activities—one of the reasons thousands of us throng to such areas for shorter or longer periods?

To return, however, to the West Coast of New Zealand, to explore perceptions further, the girls were asked to construct statements which they would use to test the way their parents and friends saw the Coast. The following three statements were voted to be the most appropriate:

1. I would like living on the West Coast. Yes/No.
2. The West Coast is an unspoiled part of New Zealand. Yes/No.
3. The West Coast has a very bright future. Yes/No.

The girls considered negative replies to all three to be the 'correct' ones, though they expected greatest disagreement on the third—the government of the day was strongly and persuasively advocating the exploitation of an almost non-renewable resource, the slowly maturing magnificent native forest, as one way of overcoming unemployment in the area.

Using slides

The final exercise was based on viewing slides taken by the girls themselves as well as some from my collection. By a process of elimination, in response to the question, 'Do you consider this slide to be typical of the Coast?' it became apparent that those slides which featured the brightly painted houses, small settlements strung out along the roads, or buildings of any kind were judged to be more typical than those featuring the high snow-capped mountains or bush-clad slopes. These gross, well-known physical features had been successfully supplemented by less spectacular details. (Compare with Long, 1953, 1961.)

Mediating public and private meanings

Discussion of this kind was, I believe, a valuable exercise in mediating and extending private and public meanings. It was certainly enjoyable and lively. We should pay more attention to private, personal meanings in the learning process as a way of developing meaning further. This may seem a utilitarian view of and role for humanistic geography. It is not. As I have already said, I believe that scientific and humanistic geography are complementary answers to the question: 'How is the nature of things to be viewed?' We all have our subjective and objective experiences of reality which may be used as data or evidence in learning activities. It is on this basis that the types of data-processing activities described in the main body of this chapter are classified.

Using one's own experience as data

The exercises now outlined should be seen along a continuum from those using data subjectively to those using data objectively. In the West Coast experiment, the students' own experience yielded the data. At the present time, cognitive mapping is probably the most developed and widely known strategy based on students' own experience and two activities fitting such a description follow.

The first gathers in data on a world scale; the second at a neighbourhood level.

Where on earth would you live?

This mental mapping exercise is designed to elicit perceptions and preferences around the globe and first probes the students' environmental knowledge of their world. This builds up a data base to be understood and explained in the first instance in relation to factors influencing such knowledge. This is followed by tasks requiring classification, analysis and interpretation.

Within the preference diagram shown in Figure 3.7, and with reference to a world map on which the forty

Figure 3.7 Where on earth would you live? A preference sorting diagram

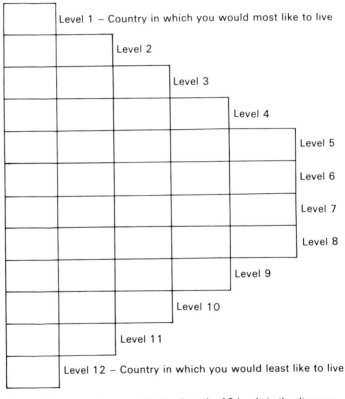

Level 1 – Country in which you would most like to live
Level 2
Level 3
Level 4
Level 5
Level 6
Level 7
Level 8
Level 9
Level 10
Level 11
Level 12 – Country in which you would least like to live

Sort the forty countries named below into the 12 levels in the diagram:

Algeria	Finland	Italy	Singapore
Argentina	France	Japan	South Africa
Australia	West Germany	Malaysia	Spain
Brazil	Greece	The Netherlands	Sweden
Canada	Hong Kong	New Zealand	Switzerland
China	Hungary	Nigeria	Tanzania
Cook Islands	India	Peru	Turkey
Cuba	Indonesia	Philippines	UK
Czechoslovakia	Iran	Poland	USA
Egypt	Israel	Saudi Arabia	USSR

countries have been named, ask a class to rank the countries 1–12 in answer to the question: 'Given a completely free choice, where would you choose to live?' It is helpful in some classes to have the countries named individually on cards so that they can be shuffled about and initial choices sorted through on a second round of reflection before being written into the diagram. Students need to appreciate that they are being forced to make one choice at the top and bottom but that there is room for more equal and/or neutral feelings in the middle.

After ranking is completed, the students should be given a chance to write a few sentences or a paragraph on the reasons for their first and last choices. The activity may then proceed so that a choropleth map of around-the-world preferences is built up, either at an individual, group or class level. I have chosen individual mapping which is then to be discussed at group and class levels. After maps have been shaded ask students to discuss the following points in groups:

1. Make a list of the reasons for your choices. (Reasons which have been given by students include (1) cultural similarities and differences, (2) climatic and scenic attractions, (3) standard of living and way of life as well as how well they *feel* they know the countries from personal or vicarious experience based on books, the news media and so on.)
2. Classify the reasons put down on your list and decide what reasons you have for most strongly and least strongly preferred countries. Do they derive from direct or indirect sources of information? Which do you think is most helpful?
3. *Where* are the countries which you have chosen as most and least preferred? Does distance appear to have any effect? Are your most/least preferred countries closest/furthest from your country? What most influences your knowledge of countries? Bring up any other ideas to explain your choices which you think the group has overlooked. Explain why you think they are important.

Finally, in order to compare preference surfaces, discussion at a class level should be promoted and the data and discussion used to apply the general ideas perhaps most usefully in the context of predicting migration. Where do most English migrants go? Why? What kinds of things are they looking for? If you were a country desperately needing English migrants what kind of an advertisement would you

write for their embassy information service? (See also Fien and Hull, 1980.)

Classroom activity

What is your neighbourhood?

Environmental knowing and meaning at the personal level can again be explored through an enquiry sequence directed at the neighbourhood scale. The activities outlined here are closely modelled on one put forward by Blachford *et al*, 1976. The enquiry sequence focuses attention on personal experience and understanding. The questions guiding the activity are as follows:

> What is your neighbourhood?
> What are the important things in your neighbourhood?
> What is the size of your neighbourhood?
> How do people's ideas of their neighbourhood vary?
> Why do people's ideas of their neighbourhood vary?
> What kinds of buildings fit in to your neighbourhood?

Neighbourhood sketch maps

The term 'neighbourhood' may require some discussion. Students could be asked to write down some ideas on a piece of jotting paper and then to draw a sketch map to show the likely neighbourhood of:

1. a baby who is not crawling;
2. a baby who is crawling but not walking;
3. a two-year-old child who is walking but not allowed on the street alone;
4. a child at primary school.

In drawing the sketch maps, the students will have to think about what kinds of places a baby or a child probably knows very well. A very young baby's neighbourhood may be only the cot and things in it like a rattle, a teddy bear and faces which peep into the cot. The general definition of neighbourhood being built up will go something like this: A neighbourhood is an area containing familiar things, people and places.

The next activity requires students to draw a sketch of their neighbourhood and to compare the things that are the same and different about the maps. The size of the area and the landmarks, buildings and streets selected can be examined and discussion focused on why the size varies and why some things are important and not others. It may be useful for the teacher to have Lynch's categories of nodes, edges, districts, landmarks and paths in mind to help guide the discussion (Lynch, 1977). A generally agreed upon class neighbourhood should be delineated. The similarities and differences in the sizes and contents of a neighbourhood will be related to interests and activity patterns and if interest in the exploration of neighbourhoods is high, then a number of group centred activities might be organised. The possibilities are numerous. Students could be asked to get the neighbourhood sketch maps of:

1. older or younger students in the school;
2. their parents;
3. elderly people;
4. a range of people in another neighbourhood;
5. people living in flats.

From such data, generalisations will be reached about people's neighbourhoods and how and why they vary in size and detail.

Towards generalisations

It is not possible to predict or set out with the same degree of precision the nature of the generalisations as in the size and spacing of settlements activity. We have moved away from geography as science and the analysis of 'objective' data to the collection of subjective data. However, generalisations will emerge at a level of the relative frequency, for example, with which (1) particular landmarks appear on sketch maps or (2) a particular street or feature seems to mark the edge of a neighbourhood.

It is the comparison of students' sketch maps with those of other people which will enable more all-encompassing generalisations to be formed. For example, it may be concluded that the neighbourhoods of people who have lived in the area for ten years or more is larger and more detailed than people who have lived there for five years or less, or people living in apartment buildings have smaller neighbourhoods than people living in single units.

Moving in neighbourhoods

Movement patterns and knowledge of an area will almost certainly have been linked in students' minds in the previous activity. Perhaps the first definition of neighbourhood as an area containing familiar objects,

people and places can be expanded to include 'and an area about which I walk, or ride often.' Answering questions such as:

How do you move around your neighbourhood?
What mode of travelling is likely to enable you to get to know your neighbourhood?
What would be the effects if you travelled everywhere by car?

help to bring out the point again.

Generally, how well a neighbourhood is known depends on the number of times a person travels along the streets and the speed of travel. The graph, Figure 3.8, illustrates the range of possibilities. A number of streets which fit each category can be chosen (refer back to known and unknown streets in the students' original maps) and the generalisations tested.

Figure 3.8 Travel through neighbourhood streets

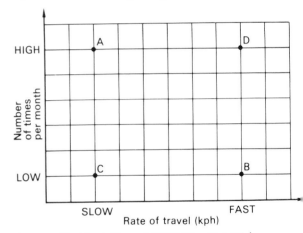

SOURCE: Blachford, K. *et al* (1976) *Way Out*, Rigby.

The neighbourhood in personal meaning

The humanistic geographer is more likely to tackle, 'What is your neighbourhood?' not as a sketch map exercise, but as an exercise in realising individual meaning or in attempting to elucidate the personal meaning of neighbourhood to students individually. Geography as personal response would want to get beyond the environment as perceived into the environment as felt experience. Activities on neighbourhoods and action space come closest to this when the liked/disliked dimension is introduced. The 'what is your neighbourhood?' activity could be extended by asking students to write statements or comments along the following lines. The exercise as set out is addressed to students. (See Farbstein and Kantrowitz, 1978.)

a) In my neighbourhood, my favourite places are . . .
In my neighbourhood, I don't like . . .
In my neighbourhood I enjoy being in a place like . . .
I feel unhappy in my neighbourhood when I'm in . . .
The most beautiful places in my neighbourhood are . . .
The ugly places in my neighbourhood are . . .
List as many characteristics as you can of places you like.
List as many characteristics as you can of places you dislike.
What parts of or things in your neighbourhood are likely to change in the next ten years?

b) Choose one of the places you enjoy and one you dislike.

c) Spend at least 30 minutes in each place. Concentrate on being in that place, forget about other things. Tune into your feelings and make a list of them. Make sure that you list your emotional reactions. Do not simply describe what is there. Then make another list of the characteristics of the place. This list can describe its physical features.

d) Try to explain why you feel as you do about each place. Is there a connection between your feelings and the physical characteristics you listed? If this is not so, what are you reacting to?

e) Compare the physical characteristics of the two places. Are they totally different or are there any similarities? Are there any conclusions you can come to about how the physical characteristics of places make you feel?

Figure 3.9 (a) Evaluating Castle Square, Sheffield

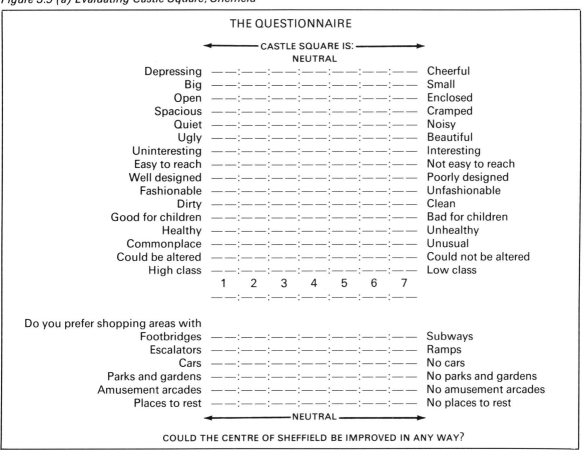

SOURCE: Tilley, R. G. (1974) 'The application of semantic differentiation in the classroom', *Profile*, Vol. 7, No. 19, Part 1.

f) If you did this project with other people compare reactions. (You may keep your results to yourself, if you wish.) What kinds of places were chosen as enjoyable? As disliked? Are there characteristics which are common to the places which most people enjoyed or disliked?

g) Can you think of any changes you could make to either of the places which would be likely to change your feelings.

Within as yet more mainstream geographical writing, the semantic differential technique is also useful for evaluating places. It provides a workable means of getting to grips with how places are experienced and responded to. Developed as part of a research study of meaning it is an attitude measurement scale, not free from academic argument and reservation, but nevertheless an expeditious technique for evaluating people's attitudes to places. For example, a town square or market place, shops, a shopping plaza or neighbourhood can be rated by using a semantic differential.

In a survey of Castle Square, Sheffield, Roger Tilley (1974) selected the bipolar adjectives listed in Figure 3.9 (a). The adjectives, separated by a seven point scale, are usually arranged so that positive and negative descriptors vary from left to right hand position. A seven point scale separates the adjectives and, in interpreting the semantic differential, a value of one to seven is assigned to each space. The space having a value of four is the neutral point. This is clear from the example in Figure 3.9 (b).

A class profile

A score of the average value assigned to each bi-polar pair of adjectives is calculated by dividing the total score by the number of students responding. In the case of Castle Square, a neutral attitude holds. For an array of adjectives, a profile can be drawn up by plotting the mean for each pair on a clean sheet and joining these together thus:

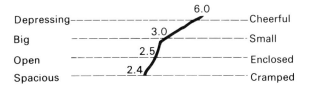

It may be that a composite overall view is not wanted and preference is given to having the students discuss their judgements. The differing views held and conflicting judgements made is a lesson in selective perception and individual preference.

As a follow-through for an exercise like this, students could be given the task of re-designing the square. The problems of re-designing to suit everyone's tastes will become obvious. Different groups in the class may be assigned roles of a planner, working to a specification laid down by a city council interested in only one aspect of the place (e.g. decreasing traffic congestion); a person working in the area; a person living there; a regular shopper (pedestrian); a regular shopper (by car); a shop owner; a tourist. The possibilities are endless and yet dependent on the place being studied. The pedestrian precinct exercise presented in Chapter 4 is an example of a suitable follow-up activity to take thinking further after an exercise like this and apply reasoning and values to real world problems or issues.

Figure 3.9 (b)

	Very	Quite	Slightly	Neutral	Slightly	Quite	Very	
Depressing —— :	—— :	—— :	—— :	—— :	—— :	—— :	—— :	Cheerful
Score	1	2	3	4	5	6	7	
Results for 20 students:	1	2	3	2	4	5	3	
Mean	1×1	2×2	3×3	2×4	4×5	5×6	$3 \times 7 = \dfrac{93}{20} = 4.06$	

Classroom activity

Evaluating landscapes

Encouraging personal responses to particular places or landscapes is possible through a variety of evaluation and appreciation exercises. Opportunities to reflect on what is liked or disliked, or what other things are associated with a place or landscape reflects another element in humanistic approaches to geographical education which link back to the landscape paradigm. Such exercises fall into the category of working with data based on the students' own experience. Students may be asked to look at a photograph of a landscape or a number of such photographs and to place each into one of the following categories:

1. Very pleasant
2. Pleasant
3. Indifferent
4. Unpleasant
5. Very unpleasant

Individual and/or class profiles could then be tabulated from 1–5 and the similarities and differences discussed. (See Wellard *et al*, 1978, p. 46.) Or students may be asked to write out a number of adjectives they would use to describe the scene. The adjectives could be chosen from an adjectival checklist if it were felt students needed some prompting. Scope for introducing poetic and other literary descriptions of landscapes and places is great in this dimension of geography.

Industrial landscapes

Peters and Larkin (1977) have reported on a particular study of industrial landscapes which could be easily replicated as an introduction to a unit of work on industry if not as a study in its own right. Students rated, as above, 20 slides of industrial landscapes and average scores were calculated. In the case of two rather different slides, one of a modern industrial park in California and the other of an older industrial scene in Milwaukee, class views were consistently opposite. The former with its landscaping and modern architectural style was pleasant, the latter, with Victorian style factory buildings, chimneys and a slag heap categorised as unpleasant.

The authors suggest discussion of reasons should be related to the following points: architectural style of buildings; building materials; windows; smokestacks; age; state of use or abandonment; landscaping of surroundings; transportation facilities and visible pollution. Landscape evaluation is, at present, an interesting growth area and may go some way to adding to the (1) observe, (2) record, tradition in field work, a third step, (3) evaluate responses.

Teachers will have their own slides and photographs to use in setting up landscape evaluation exercises. Five field sketches in BEE, 106, would also be appropriate (Hanson, 1980). The entire issue is devoted to 'Looking at Landscape'. I do not wish to give the impression that evaluating landscape is a simple or straightforward process. It is not (Gilg, 1978). But, like perception studies and semantic differential techniques, I believe vulgarisation at the classroom level in landscape evaluation is justified for the educational benefits which accrue. Good teachers are good vulgarisers.

Classroom activity

Recreational experience and satisfaction

Exploring personal levels of satisfaction in relation to preferred recreational habitats is another area which has a potential in studies based on subjective data. Helburn (1977), for example, has set up a wildness continuum along which people pursue recreational activities. At the urbanised, constructed end there are:

large numbers of people
in small areas,
close to their other activities,
for short periods
of frequent occurrence,
very much under the control of the individual,
at very low risk, and
often with much equipment and many things
deriving satisfaction.

At the wild end of the continuum there are:

relatively few people
in large areas,
usually remote from their normal daily life,
for longer periods of time,
but infrequently,
with the individual clearly not in control of the
environment,
often at considerable risk, and
normally with very little gear
deriving satisfaction.

As Helburn states, each of us has our own satisfaction curve. What is a satisfying habitat for recreation? What is wildness to you and your students? A discussion of these extreme conditions and those lying in between, backed up by slides or sketches is a potentially rewarding activity. Personal responses are being examined and compared, illuminated and evaluated. Questions could be added: for example, on preferred recreational activities and environments if costs were not a constraint. Apparent satisfaction with the local park may well relate to conditioning imposed by costs and other constraints. Teachers should not ignore chances to use geography as social criticism and map open space in cities, for example. It needs little imagination to understand how such an activity could be linked to rural land use conflicts and issues related to managing national parks (Beaver, Goodway and Rodgers, 1980; Rodgers, 1980).

Data from other people's experience

Images and feelings, preferences and opinions about places are both shared and personal: by examining what places mean to different people, our understanding of places grows and our environmental experiences, attitudes and values are raised toward a new level of consciousness. Ten opinions of the quality of city life set out in Figure 3.10, provide data from other people's experiences which might be matched to similar paragraphs written by students. The data can be analysed for the desirable and undesirable qualities of city life mentioned and the lists added to:

Extract	Desirable Qualities	Undesirable Qualities
1		
2		
3		
4		
5		
6		
7		
8		
9		
10		

Alternatively, a number of words and phrases describing city life can be picked out and the frequency of occurrence noted. Some of these phrases are: difficult to escape from; gives

opportunities to meet people; good shopping; plenty of recreational facilities; lack of communication between people; overcrowded; loneliness; noise and rush; lack of greenery.

Geography in the arts

The use and analysis of these data based on the examples of landscape appreciation and opinions on city life bring geography back into the arts side of the curriculum and obviously value-laden material is being used. These exercises require the application of intellectual abilities of comprehension, analysis and interpretation, and data can be processed and organised by rating and tabulation in a systematic fashion parallel to techniques of network analysis, for example, in scientific geography. There is, however, a literary flavour about such work and tabulation is but a preliminary to sorting out ideas through discussion and questioning. To explain the reasons for a landscape being pleasant or unpleasant requires skills in oracy or literacy, an ability to articulate one's own feelings and responses which does not, like network analysis, proceed by rules. The discussion and flow of ideas and words can often be helped by the kind of structures advocated by values educationists. Some of these procedures of values analysis and clarification are discussed in the next chapter. Only one such framework is therefore offered for discussion here.

1. Give students a scenario in which they are members of a committee choosing one of two markedly different landscape paintings for the school entrance hall. They are to make their choice on the basis of what they *like* best.
2. On a sheet of paper they are to rank the paintings 1 and 2 and to record their reasons for liking or disliking the pictures.
3. Students who ranked the same painting in first place are grouped in pairs. Each pair is to study their reasons and select what they consider to be the best two reasons for their choice.
4. Pairs having different choices then combine into fours if choices are fairly evenly divided. The two lists of reasons are explained, analysed, supported and defended.

In the clash of contrary opinions, it is intended that ideas are clarified and the reasons for opinions evaluated and developed. The nature of the generalisations will be to relate particularities and specifics to overall preferences. The extent to which

Figure 3.10 Qualities of city life

(i)
The worst part of city life is probably the aspect of getting outside the city either on business or recreation. Too many people are forced to rely on private motor cars through lack of adequate public transport. Lack of exercise and fresh air are also factors militating against city life at the present time. However, inner-city life in the future is likely to be far more pleasant than the development of suburbia with local industries and the never-ending sprawl.

(ii)
Living in a city means a mass of people who live in streets, many in one building sometimes, all living close together, and therefore all have to learn to live with each other. Trams and buses take people from place to place quickly—there is every kind of shop within walking distance to supply every want at competitive prices, so that the city dweller learns a variety of values quickly. Playing fields for children, sport arenas . . . entertainment is available at all levels and all prices. Libraries, schools, churches are usually within a kilometre.

(iii)
Really, living anywhere is all what you make it. Two homes I visit—one in a thousand square metres, nice garden, swimming pool, gracious house, music, books, good education—and friendly people who say 'good day' and smile quite nicely but no tremendously friendly situation. The other is a terrace house in an inner suburb—people all around them, which they like—situation and people. There is no real 'area problem' wherever you live—if you don't like where you are, would you be happy wherever you went? There is access to everything pretty well in a city; you can choose the way of life and enjoy it. But don't growl if you choose the wrong way. The better way is available.

(iv)
The quality of life seems reasonably good although there appears to be greater pressure over material things. City life offers greater opportunities to the individual—you experience so much more by mixing with such a variety of different people. It is stimulating. City life has an urgency about it.

(v)
City life gives you a sense of security. A feeling that you shouldn't have to worry too much about life. But there is the disadvantage of forced individualism. You are alone in the city: there is very little communication in general from one person to another. You don't really know anyone else.

(vi)
The quality of life in a city varies in relation to one's income. City life can be demanding, exciting, rewarding and exhausting. For many it is probably just a matter of existing.

(vii)
City life—like being a sardine, never the peace of being alone. Bustle—perennial rushing, noise, costume.

(viii)
'Quality of life' means different things to different people. To me it means taking *everything* into account. Do I enjoy living my day-to-day life in the city more than I would in a rural setting? Overall, I do, but I particularly resent some things—the dogged atmosphere, the dirty streets, the 'closed in' dark sections in the main city streets caused by high-rise buildings, the dirty water in the streams and seas, the noise, the lack of 'green belt' areas.

(ix)
I would describe city life to an outsider as fast, competitive, impersonal, frustrating (travel-wise) on the one hand—and exciting, challenging, colourful and fulfilling on the other.

(x)
The best part of city life is the social atmosphere with so many friends and opportunities of making more. The best place in the city is Spencer Street station for that's the place where I catch my train home. For really the city doesn't interest me except as a place to have a little fun in and a place to stay clear of. They probably would describe Spencer Street as a drab piece of working machinery, necessary only for its job of transporting people cheaply. However they, like me, could explain it as an escape, maybe?

Source: Spicer, B. J. *et al* (1974) *The Global System 4: Space for Living*, Jacaranda Wiley.

aesthetic concepts are built in depends very much on the background and interests of a teacher. (See, for example, John Wise (1977) on music and geographical education.) The landscape tradition in geography has always been strong and, to me, very central to its core and appeal.

Classroom activities

The use of photographs in scientific geography

Slides, photographs and sketches also have their place in geography as science. In the settlement siting activity elaborated in Chapter 1, sketch maps formed the data base for focusing questions and developing concepts and understanding. A linking activity may be organised around photographs used as data to answer a general question on the variety of settlement functions both within and between settlements. Photographs or landscape paintings are not used to elicit preferences and responses. Rather they are to be analysed for any evidence which hints at answers to questions such as: What are the occupations of people living here likely to be? What makes this settlement tick? What keeps it alive? On what evidence are you basing your answers? Evidence of industries and resources are part of an objective reality—sawmills, power stations, mines and wharves are part of the 'real' world which may be examined through secondary data sources like photographs. (See Slater and Spicer, 1982.)

Figures 3.11 and 3.12 are but two photographs (see Wolforth and Leigh, 1971, 1978) in a possible series which could be collected to demonstrate the

Figure 3.11 Aerial view of Britannia Beach, BC, with population and employment statistics

Source: Tourism B.C. Film Promotion.

Population and Employment in Britannia Beach, BC, 1969

Population: 709
Employment:
         Anaconda Mine and Mill
           280 employees
             90 office workers
         1 general store
         1 cafe
         1 gas station
         1 post office
         1 elementary school
         1 customs office

Source: Wolforth, J. and Leigh, R. (1971) *Urban Prospects*, McClelland and Stewart.

Figure 3.12 Aerial view of Davidson, Sask., and population and business statistics

SOURCE: National Film Board of Canada (G. H. Jarrett, 1961).

Population and Business in Davidson, Sask., 1980

Population of Davidson in 1980	1100
Population within a 25 mile radius	7000
Business and Professional Services	

Agriculture	*No.*
Nurseries	1

Construction	
General building contractors	1
Plumbing and heating	2
Painting and decorating	1
Electrical	3
Plastering and drywall	1
Others	1

Manufacturing	
Printing, publishing	1

Motor Freight Transport	
Local trucking firms	1
Motor freight transport	4

Retail trade	*No.*
Bulk oil	3
Lumber yards	2
Hardware stores	2
Farm equipment	5
General merchandise	1
Grocery stores	1
Bakeries	1
Motor vehicle dealers	3
Auto supply stores	2
Service stations	5
Family clothing	2
Furniture stores	1
Eating places	6
Drug stores	1

Services—personal	
Coin laundries	1
Photo studios	1
Beauty shops	3
Barber shops	1
Funeral parlours	1

Auto services	*No.*
Auto body repair shops	2
Auto repair	3
Car washes	1

Repair shops	1

Entertainment and recreation	
Movie theatres	1
Bowling alleys	1
Public golf courses	1

Services—financial	
Credit union	1
Bank	1
Real estate	4

Services—professional	
Dentist	1
Doctor	1

Accommodation	
Hotel	1
Motel	1

SOURCE: Saskatchewan Industry Department.

variety of single-function and multi-function settlements arising from the earth's variable resource base and man's changing needs and appraisals.

A mining settlement

Figure 3.11 is an aerial view of Britannia Beach, in British Columbia, Canada, taken before it was closed down in the 1970s. Students are likely to suggest a number of hypotheses for the main function of the settlement including perhaps saw-milling, mining of some kind, or hydro-power production. Other jobs besides those associated with the main activity doubtless exist and suggestions will cover retailing and service functions as well as educational and government services.

The photograph does not provide certain evidence on any of these points but opens up a discussion of possibilities. What other evidence is needed? Statistics on the population and employment structure of Britannia Beach, here accompanying the photograph, put an end to speculation. The factual evidence afforded by the photograph highlights a significant constraint which must be recognised in the use of both maps and photographs. Statistical and other documentary evidence is usually essential. Photographs often raise more questions than they answer but nevertheless they are interest-arousing data sources for geography lessons and it is important for students to be made aware of the dangers of relying unduly on one piece of evidence. I would adapt the Liverpool Project's table on evidence to read thus:

| Recognising and interpreting evidence | → Drawing conclusions from evidence | { Is the source of evidence reliable, up to date? *Is it the only source of evidence available?* Is the evidence presented from one point of view? Is the conclusion the only possible one? |

SOURCE: Based on Blyth W.A.L *et al* (1976) *Curriculum Planning in History, Geography and Social Science*, Schools Council/Collins.

An agricultural centre

Hunches on the function of the town in Figure 3.12 may not be so diverse though that is really dependent on students' backgrounds. Davidson is a service centre in an agricultural region and detail on the population and business activities of the town is provided. Reaching tentative conclusions through examining one piece of data is a useful exercise when other data can be supplemented to further clarify suggestions and ideas. More evidence is being brought in, not to challenge already formed generalisations, but to assist in the process.

The use of statistics

The exploration of settlement functions through photographs and statistics in particular, can be taken a little further and related back to the size and spacing of settlements in the Shepparton district (refer back to Chapter 2). Figure 3.13 sets out the retail functions to be found in each of the forty-one settlements. These data can be used to establish a number of generalisations about centrality and the urban hierarchy. An appropriate question sequence is:

Is there a hierarchy of urban places?
Is there a hierarchy of functions?
What functions characterise small settlements?
Which functions can be identified as low order/ high order?
What is the threshold population associated with particular functions?
Which functions are least centralised/most centralised?

Population and functions

Inspection of Figure 3.13 makes clear a number of well-known relationships such as that between population and retail functions. A graph would serve as a means of further processing the data and presenting them in visual form. The varying numbers of different retail activities plotted successively against population would reveal the functional hierarchy and identify high and low order functions across the forty-one settlements. A class would need to be organised so that each person or pair did one function and results were then pooled. Once visually displayed the question of threshold populations becomes more obvious. A little show of numeracy, rather than words and literacy, establishes the threshold values and may be taken as an example of arithmetic in the service of geography and in the processing of data.

Settlement functions

If a town with 1,000 people has one supermarket, another with 2,000 people, two and another with 3,000, three, then we can conclude that 1,000 people are needed to support one supermarket. In the Shepparton district, assuming that this is a unified and self-contained retail area, the smallest settlement, Nalinga (40 people) supports a motor service supplier while the next town with any functions, Youanmite (100 people) supports a motor service supplier and a food supplier. Garages would seem to be the lowest order service, followed by food shops. Sometimes the threshold population is calculated in a slightly more sophisticated way than by merely inspecting a table and matching population with the first occurrence of a service.

Calculating threshold values

To do this, add the population of all settlements— from the smallest settlement where the function appears up to the settlement which is one above the largest *without* that function. Add together only the population figures of the settlements possessing the function and divide the total by the number of times the function is present. The threshold values calculated in this way for each of the functions performed by settlements in the Shepparton district have been incorporated into Figure 3.13. The threshold values of the functions can then be ranked and low and high order goods or services differentiated.

A new industry

A planning problem would be appropriate as a follow up. For example, a group of farmers in the Nalinga area have been persuaded to grow quantities of a variety of sugar beet suited to the district and to be used as a new source of energy. A factory employing 200 people, including chemists and research workers, is to be built in Earlston. Each employee can be presumed to bring a wife and two children to the town. Since the project is a high prestige one likely to attract international visitors, the government has decided to take the trouble to plan a shopping area. You are one of two local representatives on the planning board and the only one with any geographical training. The board is discussing the problem of what shops (and how many) are needed. Set out your choices and your justifications in a neat, one-page report which you have had the chairperson circulate to all board members.

Calculating a centrality index

The idea of centrality like threshold can be given more precision through some numerical work and calculating a centrality index is another example of the usefulness of simple numerical methods in data processing. The identification of centrality by means other than size of population and number of functions can be reached through the calculation of a centrality index (W. Davies, 1967; G. Davies, 1978). The centrality index provides a measure of the relative significance of a town's functions and its standing in a group of settlements in an area. Centrality is measured by the extent to which each function is dispersed throughout a group of settlements. Satisfaction of demand is considered to be spread more widely the more dispersed the function. Large coefficients indicate more centralised functions. The measure of centrality, a location coefficient, is calculated by the formula:

$$c = \frac{t}{T} \cdot 100$$

where C = the location coefficient of function t, t = one outlet of the function t, and T = total number of outlets of function t throughout the area being studied.

The *coefficient* for accountancy firms in the Shepparton area is $\frac{1}{10} \times 100 = 10$ and the *centrality value* for this function in Shepparton is $4 \times 10 = 40$ or the total number of accountants in Shepparton (4) multiplied by the coefficient for accountants (10). Figure 3.13 also shows, for the sake of interest, the centrality values for the range of functions found in Shepparton. Totalled, these values give a centrality index of 848 to Shepparton. Can we expect it to have the highest centrality index? Are the centrality values what would be expected? Rank the services according to their centrality indices.

This is a rather elementary example of statistical data processing given the last ten to twenty years of numerical bombardment of the subject. However, there are numerous references now available and obviously, the general purpose of this book on activity planning is to offer examples of a variety of forms of data and data processing techniques.

Maps as data

While centrality has been examined by the calculation of coefficients, a closely related idea, isolation, can now be examined via a map and some tabulation.

Figure 3.13 Retail functions in forty-one settlements

Functions and no. per settlement	19,409 Shepparton	5,806 Kyabram	3,532 Mooroopna	3,195 Cobram	2,581 Numurkah	2,510 Tatura	1,278 Nathalia	1,071 Rushworth	899 Tongala	800 Invergorden	780 Yarroweyah	760 Katunga	640 Ardmona	570 Toolamba	564 Stanhope	522 Murchison	496 Merrigum	382 Strathmerton	340 Undera	315 Tallygaroopna
Accountants	4	1		2	2	1														
Agricultural machinery shops	11	10	1	10	5	4	6		2			1			2	2				
Banks	13	10	4	6	10	8	5	3	4						3	3	4	3		2
Builders	21	10	4	9	8	7	4	3				1			1	3	2			
Carriers	26	9	3	7	3	11	4	1	1	4		1	2	2	2	2	3	3		2
Chemists	9	3	1	2	2	5	1	1	1							1			1	
Clothing shops	29	17	3	10	6	6	4	2	3						2	2				
Courthouses	1			1																
Dentists	5	1			1															
Doctors	9	4	1	2	3	1	1	1	1							2				
Dry cleaners	5	3	1	1	2	1	1	1												
Electrical shops	11	3	1	5	3		1	2	1						2	3				
Food shops	54	27	11	13	17	17	8	8	10	1	2	3	1	1	9	4	6	6	2	2
Furniture shops	10	2	2	2	4	1	3	1							2					
Gift shops, jewellers	6	4		5	5	2	3		1						1	2	1			
Hairdressers	20	6	3	7	7	5	2	3	2						2					
Hardware shops	3	5	1	6	2	2	5	2	4						3	1				
Hotels	6	3	3	3	3	3	3	3	1					1	1	3		1	1	1
Motor services	30	10	4	9	9	7	7	3	2	1	1	1	1		3	3	2	2	1	1
Musical shops	3				1															
Newsagents	5	1	1	1	2	1	1	1	1			1			1				1	
Post offices	3	1	1	1	1	1	1	1	1	1	1	1			1	1	1	1	1	1
Solicitors	6	2			2	1	1	1												
Sports goods shops	11		1		5	3	3		1						1					
Stock markets and stations	10	7	2	6	3										1	1				
Total functions per settlement	316	139	48	108	106	87	64	37	36	7	4	9	4	6	37	33	19	16	7	9
Centrality index																				

310 Waara	308 Giragerre	300 Katamatite	300 Dhurringile	280 Lancaster	280 Lemnos	260 Koyuga	260 Wyuna	240 Barmah	200 Picola	200 Pine Lodge	190 Toolamba West	160 Wunghnu	130 Byrneside	100 Youanmite	80 Caniambo	70 Kanyapella	60 Baileston	60 Hendersyde	40 Nalinga	40 Earlston	Centrality Value Shepparton	Centrality Coefficient	Totals of each function for the area	Threshold values
																					40	10	10	2348
	1																				20	1.82	55	333
1	2	2		1					3												15	1.15	87	163
																					29	1.37	73	341
	2			1					4	1		2									27	1.04	96	199
																					33	3.7	27	587
1				1																	34	1.16	86	288
																					50	50	2	11302
																					71	14.3	7	4193
																					36	4.0	25	473
																					33	6.6	15	1071
		1																			33	3.0	33	555
2	3	5		1	1		1	1	3			3	1	1							24	.45	224	114
																					37	3.7	27	545
																					21	3.45	29	692
																					34	1.72	58	391
																					20	2.56	39	248
1				1		1		1	1												14	2.4	41	350
1				1		1		1				1		1					1		28	95	105	243
																					75	25	4	5497
1	1			1								1									23	4.55	22	465
1	1	1	1	1		1	1	1	1	1	1	1	1	1	1	1	1	1	1	1	7	2.4	41	269
																					50	8.3	12	2029
																					42	3.85	26	1216
																2					21	3.13	32	490
8	10	12	1	6	1	3	2	4	12	2	1	8	2	3	1	3	1	1	2	1	31			
																					848			

Figure 3.14 Meteghan

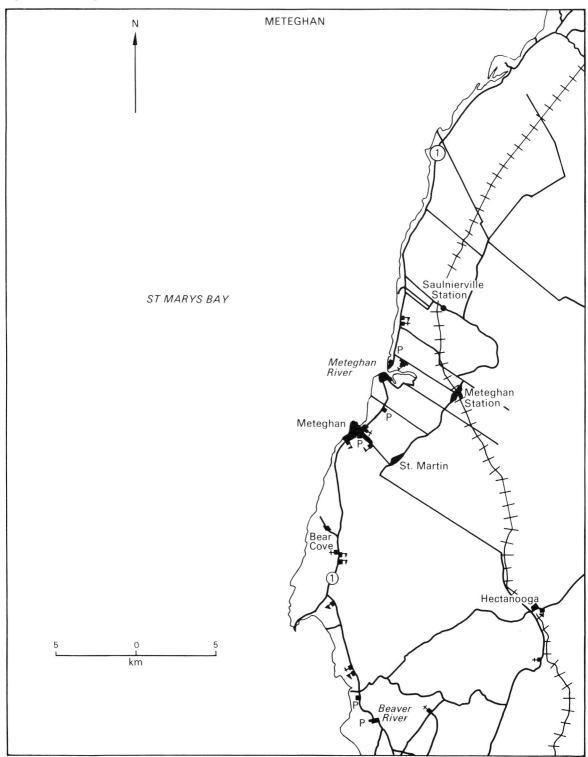

SOURCE: Day, D. and Millward, H. (1977) 'Analysis of village siting and rural isolation from topographic maps', *Geoscope*, Vol. 10, No. 2.

Unless teachers have taken photographs for a precise purpose to illustrate particular points, maps probably provide a richer base for developing ideas and drawing conclusions. The clarity and legibility of maps often gives them some advantages.

Rural isolation

A group of Nova Scotian educators (Day and Millward, 1977) involved in curriculum renewal devised an interesting exercise to examine the concept of rural isolation and to develop a general measure of isolation. The exercise is another example of data processing to elucidate concepts and general ideas and it balances nicely with the work on centrality. What is rural isolation? How can we measure isolation? What criteria do we select? How do changes affect rural isolation? What are the effects of isolation?

The exercise is based on the 1:50,000 topographic map of the Meteghan area (21 B/1 Second Edition) located along the shores of St Mary's Bay in Digby County, Nova Scotia. The area, originally settled by the French, is marked today by two lines of settlement: one adjacent to the coast, the other further inland (Figure 3.14). Eight settlements are selected for study: Bear Cove; Beaver River; Hectanooga; Meteghan; Meteghan River; Metaghan Station; Saulnierville Station and St Martins. In making measurements from the topographic map, it is assumed that the town centres equate with the grid references given for each settlement in Figure 3.15. Relative isolation is defined by (1) road distance in kilometres from the centre of each of the settlements to the nearest post office, (2) the shortest road distance from the centre of each settlement to Highway 1, (3) the shortest road distance from the centre of each settlement to the nearest railway station, (4) the shortest road distance to the nearest

elementary school, and (5) the shortest road distance to a high school. There is only one high school within the map area (grid reference 291008). It is assumed that the other schools shown cater for the full range of elementary schooling up to high school. Which facets of isolation can be measured, which cannot?

An index of isolation

However, given the above criteria and a completed Figure 3.16, an index of isolation can be calculated for each settlement. Each measure of isolation is taken to be of equal value and points given to each settlement based on distance from each amenity as set out in Figure 3.16. The lower the total score, the greater is the isolation. The most and least isolated settlements can now be identified, likewise the factors contributing most to relative isolation. The equal weighting given to the measures of isolation may not be acceptable and a class discussion may conclude that accessibility to one or two of the measures is twice as important as accessibility to others. If this is the case then the scores can be suitably re-weighted and a second total calculated. What significant differences occur between the weighted and unweighted indices? A further variable, perhaps accessibility to a church, could be built into the index or a class may wish to suggest another factor. Figure 3.16 gives the necessary information on distances.

Decreasing isolation

An alternative to building in further criteria is to posit, for example, the building of a new high school at Salmon River. What impact would this have on the relative isolation of the settlements? Or perhaps several elementary schools are to be closed because of falling rolls. How does this affect the relative isolation of the settlements? Given this scenario, what are the effects of isolation? What other dimensions of isolation are there?

Figure 3.15 Shortest road distance (in km.) to settlements

Settlement	Post office	Highway 1	Railway station	Elementary school	High school
Bear Cove (252907)	4.8	0.7	15.5		
Beaver River (284796)	0.0	0.0	14.5		
Hectanooga (378863)	12.6	12.5	0.0		
Meteghan (268965)	0.2	0.0	8.5		
Meteghan River (283002)	0.0	0.0	5.5		
Meteghan Station (325989)	5.0	4.0	0.0		
Saulnierville Station (315030)	7.5	2.0	0.0		
St Martin (291948)	3.2	3.0	5.5		

Source: Day, D. and Millward, H. (1977) 'Analysis of village siting and rural isolation from topographic maps', *Geoscope*, Vol. 10, No. 2.

Figure 3.16 Settlement scores on selected measures of isolation

Settlement	Post office	Points awarded for nearness to				Unweighted index of isolation (total points)	Weighted index of isolation (A)	Distance to church (in km)	Points for nearness to church	Weighted index of isolation (B)
		Highway 1	Railway station	Elementary school	High school					
Bear Cove								2.5		
Beaver River								4.0		
Hectanooga								0.2		
Meteghan								0.2		
Meteghan River								2.5		
Meteghan Station								6.5		
Saulnierville Station								4.8		
St Martin								4.0		

SOURCE: Day, D. and Millward, H. (1977) 'Analysis of village siting and rural isolation from topographic maps', *Geoscope*, Vol. 10, No. 2.

Maps in model making

The significance of maps as data bases and their role in developing concepts and generalisations is already plain from the map based activities described. One further example, using maps, illustrates educational aims and methods of data processing well within the scientific tradition in geography as map information is analysed to build a model of a British National Park. Many map exercises have been devised to test agricultural models and urban land-use models (model-making activities are less plentiful but perhaps more valuable). After all, if models have a number of simplifying assumptions built into them, then it is not surprising to find that reality departs from the model; to criticise the model on those grounds is not to appreciate its internal validity. To begin the exercise by building a model helps to dispel that confusion perhaps.

A National Park model

Goring (1977) felt that the National Park exercise he devised included the following skills and techniques which are part of the requirements set out by the Scottish Certificate of Education Examination Alternative Higher Syllabus:

1. The use of techniques of analysis to aid the understanding of properties and distribution of geographical phenomena;
2. The construction of matrices to store selected information;

3. The realisation that models are a simplification of complex reality and that at the end of the activity students would have more than a superficial knowledge of 'their spatial distribution, their characteristics and their function.'

Five maps form the basic evidence for the exercise: 1:63360 North York Moors OS Tourist Sheet; 1:63360 Peak District OS Tourist Sheet; 1:63360 Lake District OS Tourist Sheet; 1:63360 Dartmoor OS Tourist Sheet; 1:250,000 Wales and The Marches OS Tourist Sheet (including Pembroke Coast National Park).

Preparing sketch maps

Students are directed to prepare sketch maps of each Park to show relief, drainage, communication patterns, settlement and any features related to leisure and recreational activities. The information was then organised into a matrix. Figure 3.17 illustrates the general matrix as well as the sketch and matrix for the North York Moors.

A composite matrix

The five maps and matrices were then used to construct a composite matrix on which the frequency of occurrence of items in the Parks was recorded. A representative matrix was derived from this and a National Park model sketched out. These steps are illustrated in Figure 3.18. The model making has been achieved and while Goring suggests testing it against other National Parks, I would feel that as an insight into model building the exercise need not be taken further. Considerable discussion time could be spent enunciating the rules for the model's construction and suggesting how initial observations and rules could be seen as assumptions. Here is a chance to contribute to students' understanding of what is involved in the process of model making and not merely to learn or test the content of the model.

Figure 3.17 (a) General and specific matrices for building a National Park model

NAME OF NATIONAL PARK			Just outside National Park	On periphery of National Park	In centre of National Park
1. PHYSICAL FACTORS:	Yes	No			
a Is the National Park an upland area compared to the surrounding country?					
b Is the National Park a compact unit?					
c Are water - sites available?					
2. HUMAN FACTORS :					
a1 Is the National Park served by motorway?					
a2 Is the National Park served by "A" class roads?					
a3 Is the National Park internally connected by "B" class roads ?					
b1 Is there hotel/inn accommodation?					
b2 Are there camping/caravanning sites?					
b3 Are there Youth Hostels?					
c1 Are there Mountain Rescue Stations?					
c2 Are there any viewpoints?					
c3 Are there any Information Centres?					

Figure 3.17 (b)

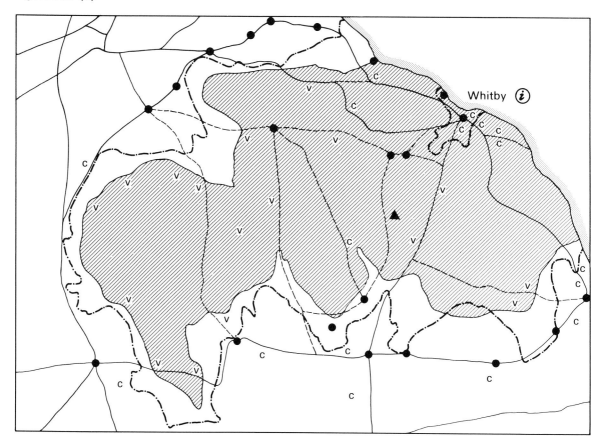

Whitby

NORTH YORK MOORS NATIONAL PARK

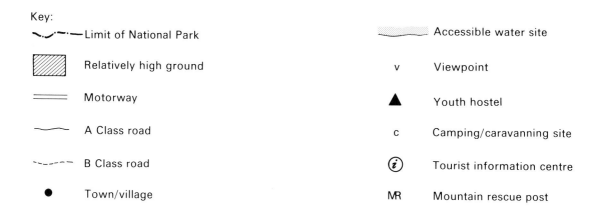

Key:

～·⌣·¬·— Limit of National Park	▨▨▨ Accessible water site
▨ Relatively high ground	v Viewpoint
═══ Motorway	▲ Youth hostel
～～ A Class road	c Camping/caravanning site
⁻·⁻·⁻· B Class road	ⓘ Tourist information centre
● Town/village	MR Mountain rescue post

Figure 3.17 (c)

1.	Yes	No	Just outside National Park	On periphery of National Park	In centre of National Park
a	*				*
b	*				
c	*			*	
2. a1		*			
a2	*		*	*	
a3	*			*	*
b1	*		*	*	
b2	*		*	*	
b3	*				*
c1		*			
c2	*			*	*
c3	*			*	

EXAMPLE: NORTH YORK MOORS NATIONAL PARK

(* indicates the presence of this feature on the O.S. map)

SOURCE: Goring, R. T. (1977) 'A British National Park Model', *Sagt*, Journal of the Scottish Association of Geography Teachers, No. 6.

Models for experimentation

Scientific geography uses many types of models, hardware models, analogue models and mathematical models to mention but three common categories. The use of models yields many opportunities for data processing and interpretation. The effects of a mix of variables in a model over a period of time gives opportunities for observing patterns and reaching generalisations. (See also Slater, 1974.) (Bartlett (1982) has made an analysis of systems thinking and teaching strategies, relevant to the discussion in this chapter.)

The map in Figure 3.19 is an area of generally

Figure 3.18(a) A British National Park model

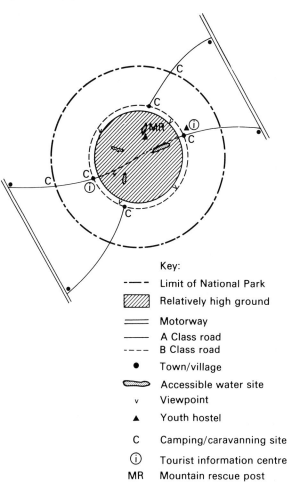

Key:
- – – – – Limit of National Park
- ▨ Relatively high ground
- ═══ Motorway
- ───── A Class road
- - - - - B Class road
- ● Town/village
- ⬭ Accessible water site
- v Viewpoint
- ▲ Youth hostel
- C Camping/caravanning site
- ⓘ Tourist information centre
- MR Mountain rescue post

SOURCE: Goring, R. T. (1977) 'A British National Park Model', *Sagt*, Journal of the Scottish Association of Geography Teachers, No. 6.

scattered population with four clusters of above average density. To simulate the growth of the settlement pattern, Walker (1979) specifies a number of conditions under which the simulation is to take place. The first of these are:

1. Areas with or near higher population densities are likely to attract settlement;
2. Settlement is more likely to take place on flat land;
3. Settlement is more likely to occur on good agricultural land rather than poor agricultural land;
4. Land near transport routes is likely to be used for settlement;

Figure 3.18(b) A measure of how well the model fits reality

RECORDED FEATURES / NATIONAL PARK	Upland 1 a	Compactness b	Water-sites c	Motorways 2 a1	A class roads a2	B class roads a3	Hotels/inns b1	Camping/caravanning b2	Youth Hostels b3	Mountain rescue c1	Viewpoints c2	Tourist information c3
NORTH YORK MOORS	*	*	*c		*	*	*	*	*		*	*
DARTMOOR	*	*	*	*	*	*	*	*	*		*	*
EXMOOR	*	*	*c	*		*	*	*				*
SNOWDONIA +	*	*	*		*		*	na	*	*	*	*
LAKE DISTRICT	*	*	*	*	*	*	*	*	*	*	*	*
PEAK DISTRICT	*	*	*	*	*	*	*	*	*	*	*	*
PEMBROKE COAST + +			*c		*	*	*	na	*		na	na
BRECON BEACONS + +	*	*	*	*	*	*	*	na	*		na	na

Key: c = Coastal water sites
+ = From 1:126,720 map
+ + = From 1:250,000 map
na = Not available because of map scale

SOURCE: Goring, R. T. (1977) 'A British National Park Model', Sagt, Journal of the Scottish Association of Geography Teachers, No. 6.

5. Land near a coalfield is likely to be used for settlement;
6. Marshy land is likely to be unattractive for settlement;
7. Coastal land with its aesthetic appeal is more likely to attract settlement.

These generalisations determine the weightings assigned to grid squares on the converted base map (Figure 3.20) thus:

1. Areas with higher than average population density, 6; adjacent cells, 4;
2. Flat land, 3; upland areas, 1;
3. Good agricultural land, 2; poor agricultural land, 0;
4. Grid square with transport route, 3;
5. Coalfield squares, 4;
6. Cells with marsh, −3;
7. Cells adjacent to coast, 1.

In this model, the base map and grid squares now become a matrix of weightings based on the above specifications. The weightings on each variable for each cell are recorded and the weightings summed up as illustrated in Figure 3.21. The element of experimentation is now built into the model by making the patterns generated dependent on the drawing of random numbers. Cells are allotted a sequence of numbers proportional to their weightings and a simulation matrix set out as in Figure 3.22.

Simulating settlement

The settlement pattern is simulated by drawing random numbers. Figure 3.23 shows the growth based on 85 drawings. Walker's suggestion to compare this with Christaller's model is perhaps unfair, given the differences in assumptions between

Figure 3.19 Original area before settlement pattern emerged

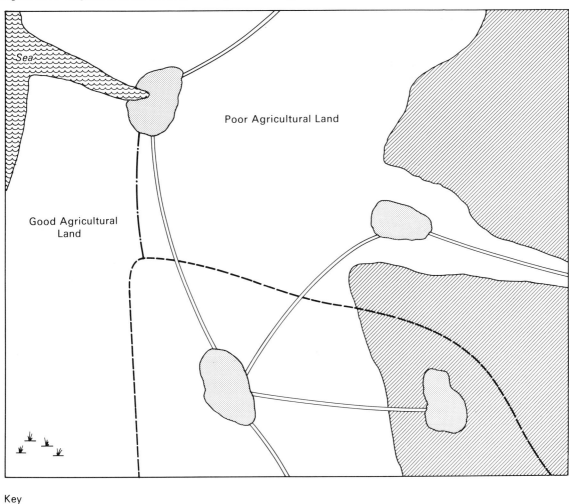

Sea

Poor Agricultural Land

Good Agricultural Land

Key

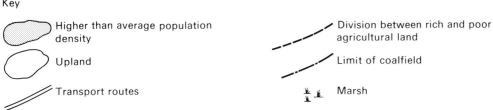

Higher than average population density

Upland

Transport routes

Division between rich and poor agricultural land

Limit of coalfield

Marsh

SOURCE: A. H. Walker (1979) 'Monte Carlo Simulation as a teaching technique', *Geoscope*, Vol. 13, No. 1.

the models. However, the concepts associated with central place theory provide plenty of material for discussion and predicted outcomes based on the rules can be compared with actual outcomes. A simulation like this provides learning opportunities similar to those achieved in observing deltas being built up in a stream table. Like the size of the grains and force of the water in a stream table, rules and circumstances can be changed to test hypotheses and ideas. Just as significantly for learning, students are engaging in determining and operating rules in a multi-variate world. The experiment, in more than one sense, certainly places students in the role of data processors and interpreters.

Figure 3.20 Base map converted to grid cells

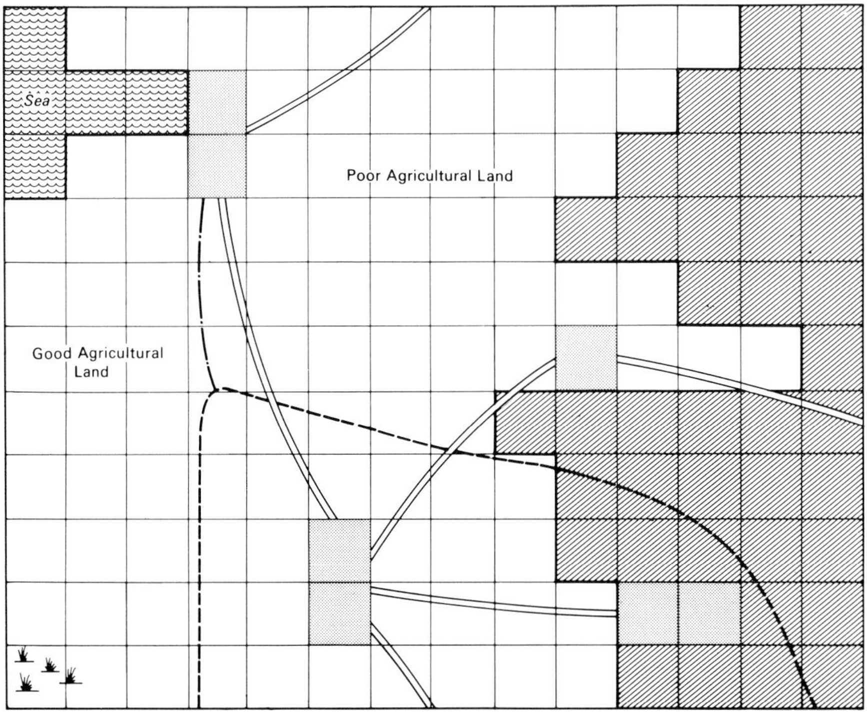

Sea

Poor Agricultural Land

Good Agricultural Land

SOURCE: A. H. Walker (1979) 'Monte Carlo Simulation as a teaching technique', *Geoscope*, Vol. 13, No. 1.

Figure 3.21 Cell weightings

13	*143*	*43*	*43*	*33*	*33*	*3*	*3*	*3*	*3*	*3*	*1*	*1*	
4	8	7	7	6	6	3	3	3	3	3	1	1	
		633	*433*	*33*	*3*	*3*	*3*	*3*	*3*	*1*	*1*	*1*	
		12	10	6	3	3	3	3	3	1	1	1	
	123	*243*	*633*	*43*	*3*	*3*	*3*	*3*	*3*	*1*	*1*	*1*	
	6	9	12	7	3	3	3	3	3	1	1	1	
123	*23*	*234*	*433*	*34*	*3*	*3*	*3*	*3*	*1*	*1*	*1*	*1*	
6	5	9	10	7	3	3	3	3	1	1	1	1	
23	*23*	*23*	*33*	*33*	*3*	*3*	*3*	*43*	*43*	*43*	*1*	*1*	
5	5	5	6	6	3	3	3	7	7	7	1	1	
23	*23*	*23*	*3*	*33*	*3*	*3*	*3*	*433*	*633*	*433*	*33*	*33*	*1*
5	5	5	3	6	3	3	3	10	12	10	6	6	1
23	*23*	*23*	*43*	*334*	*43*	*3*	*33*	*413*	*41*	*41*	*1*	*1*	*13*
5	5	5	7	10	7	3	6	8	5	5	1	1	4
23	*23*	*23*	*43*	*443*	*4433*	*4433*	*334*	*43*	*41*	*1*	*1*	*1*	*1*
5	5	5	7	11	14	14	10	7	5	1	1	1	1
23	*23*	*23*	*43*	*443*	*6343*	*4433*	*43*	*43*	*441*	*441*	*441*	*41*	*1*
5	5	5	7	11	16	14	7	7	9	9	9	5	1
23	*23*	*23*	*43*	*443*	*6343*	*4433*	*433*	*433*	*4433*	*6431*	*6431*	*44*	*1*
5	5	5	7	11	16	14	10	10	14	14	14	8	1
-3 3	*23*	*23*	*43*	*443*	*433*	*4433*	*43*	*43*	*443*	*441*	*441*	*441*	*1*
0	5	5	7	11	11	14	7	7	11	9	9	9	1

SOURCE: A. H. Walker (1979) 'Monte Carlo Simulation as a teaching technique', *Geoscope*, Vol. 13, No. 1.

Figure 3.22 Simulation matrix

	4	5–12	13–19	20–26	27–32	33–38	39–41	42–44	45–48	49–51	52–54	55	56
			57–68	69–78	79–84	85–87	88–90	91–93	94–96	97–100	101	102	103
	104–109	110–118	119–130	131–137	138–140	141–143	144–146	147–149	150–152	153	154	155	156
157–162	163–167	168–176	177–186	187–193	194–196	197–200	201–204	205–207	208	209	210	211	212
213–217	218–222	223–227	228–233	234–239	240–243	244–246	247–249	250–256	257–263	264–270	271	272	273
274–278	279–283	284–288	289–291	292–297	298–300	301–303	304–306	307–316	317–328	329–338	339–342	343–348	349
350–354	355–359	360–364	365–371	372–381	382–386	387–389	390–395	396–403	404–408	409–413	414	415	416–419
420–424	425–429	430–434	435–441	442–452	453–466	467–470	471–480	481–487	488–492	493	494	495	496
497–501	502–506	507–511	512–518	519–529	530–545	546–559	560–566	567–573	574–582	583–591	592–600	601–605	606
607–611	612–616	617–621	622–628	629–639	640–655	656–669	670–679	680–689	690–703	704–717	718–731	732–739	740
	741–745	746–750	751–754	755–768	769–779	780–793	794–800	801–807	808–818	819–827	828–836	837–845	846

SOURCE: A. H. Walker (1979) 'Monte Carlo Simulation as a teaching technique', *Geoscope*, Vol. 13, No. 1.

Figure 3.23 Simulated settlement pattern

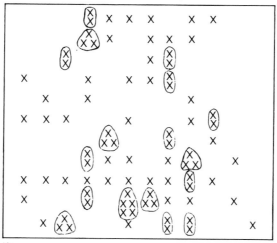

Key

X Village   (cell with one random number)

ⓧⓧ Town   (cell with two random numbers)

Large Town   (cell with three random numbers)

City   (cell with four or more random numbers)

SOURCE: A. H. Walker (1979) 'Monte Carlo Simulation as a teaching technique', *Geoscope*, Vol. 13, No. 1.

Multi-media activities

The logical capstone to this chapter is to provide a description of an activity which uses a combination of data forms and data processing methods. This would help to reinforce the point that geography teachers have a rich and varied range of resources to use in planning learning activities. The number of packs and kits of material now available, makes the selection of any one an invidious task. The revised edition of the American HSGP units (1979) uses a full range of resources—readings, photographs, maps, recordings, models—learning activities and teaching strategies in a well integrated fashion. The British Schools Council GYSL project is another attractively produced set of materials combining an array of data forms in its units (but note certain criticisms of it now being made from a multicultural perspective (Gill, 1981)). Video tapes have been especially made to highlight some themes, e.g. 'It's ours whatever they say.' Materials from the Bristol project are similarly varied with cartoons and topical newspaper extracts providing stimulation. The underground press is replete with Bristol-inspired assessment exercises which may also be used as valuable springboard or ignition resources. (For published examples see BEE, October 1974.) Planning games such as Wantage (Rawling and Rawling, 1979) integrate a variety of media forms—slides, letters, maps—and activities—field reconnaissance, role playing, and a public meeting—to explore a problem and reach a decision. The teacher as a manager of resource-based learning, operating in the interactionist style as defined and popularised by the Bristol team (Tolley and Reynolds, 1977) is realised in resource-rich, carefully prepared activities.

Classroom activity

Planning for Western Port

I have chosen to outline in a little more detail, the *Planning for Western Port* activity developed by Spicer (1974). It is a very adequate example of a multi-media activity directed towards reaching general understandings and decisions through data processing and interpretation and it introduces readers to the value laden nature of geographical education explored at length in the next chapter.

Information is presented in various forms, and students individually and in groups attempt to produce a plan on large outline maps provided for the future development of the Western Port region adjacent to Melbourne.

The activity opens with a news broadcast which clearly and concisely sets up the dimensions of the planning problem as well as posing the question of how the region might be developed into the wider economic context of society's prevailing economic and environmental concerns, priorities and values. How can Western Port's competing attractions for industrialists and holiday makers, farmers and fishermen, conservationists and shop keepers be reconciled in a planned development? What constitutes wise decision making for the Western Port region? A series of twenty slides orients viewers to Western Port landscapes and a booklet provides solid background information on all aspects of the region.

Opinions on Western Port's development

Mrs Margaret Ross, a member of the Save our Wild Life Association, states that

> our concern for materialistic things must be reduced. Why should having a great industrial region be regarded more highly than having areas close to the city with an abundance of wild life, natural vegetation, open space and fresh air? We need to get our priorities straight.

An industrial executive, Mr Bill Speel, argues differently. He emphasises enormous benefits and additional jobs and clearly sees Western Port as an ideal site for industrialisation. It has a deep water harbour, transport links, proximity to the Melbourne market, labour supply and so on.

Residents of the Western Port region believe that it is vital to inform government planners of their views. A series of meetings is planned to produce a set of principles on the future form of development. Role cards are distributed and students assigned to 'interest groups', A–H. These represent the following groups:

A Primary industry
B Secondary industry
C Conservationists
D Business and professional people
E Phillip Island Action Committee
F Hastings's Council Development Sub-Committee
G Recreational interests
H Land Development Group

Interest groups meet

The interest groups then hold independent meetings. They endeavour to reach a consensus among themselves concerning their hopes and ideas for the future of the area. Each participant must put his or her own viewpoint. Each interest group then appoints a leader to state the views of the group at the first public meeting. It is suggested that minutes should be kept of each meeting.

The first public meeting

The first public meeting has been called by the Hastings's Council Development Sub-Committee with the intention of setting up 'planning bodies'. Each interest group seeks membership of the planning bodies. At the first public meeting it is suggested that minutes should be kept by the teacher and circulated to all participants. At this meeting each interest group, through its leader, airs its views publicly and this is followed by a period of general comment, debate and questioning. The meeting ends with the formation of several planning bodies made up of a mix of interest groups.

A planning body has a composition resembling the following: a farmer, a steel industry representative, a resident of Phillip Island, a bank manager, a local school teacher, members of the Phillip Island Action Committee and the Hastings's Council Sub-Committee, a walking club representative and a member of the Land Development Group.

The interest groups then meet again and decide what their views will be as working members of the planning bodies. A period for individual study of activities, slides, the map and further information available from newspapers, letters or articles, is suggested at this point.

Planning bodies meet for up to 6–10 hours to produce a set of guiding principles and a plan for the development of the region. The principles and plan should, it is suggested, refer to all aspects of development including residential, industrial, commercial, educational, recreational, transport, conservation and pollution matters. Each plan should include a land-use zoning map.

The second public meeting

At the second public meeting, with the teacher again keeping minutes, the planning body leaders present the group plans. He or she will describe the plan and answer questions from the floor. After general

Figure 3.24 Western Port Bay, Victoria, Australia

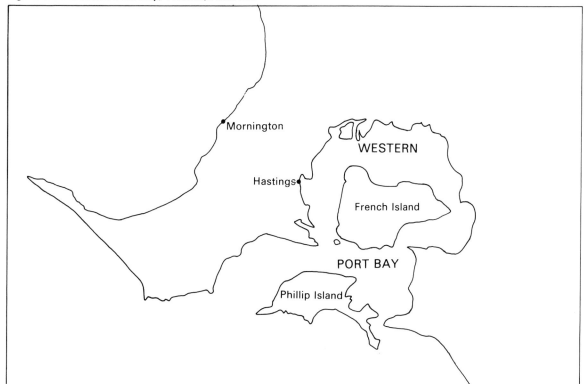

Mornington

WESTERN

Hastings

French Island

PORT BAY

Phillip Island

discussion a vote is taken to select the plan which is to be submitted to the government. Participants are not bound to vote for the plan of their own planning body. The plan selected is then subjected to modifications based on suggestions made by participants. The suggestions are to be put forward as motions and voted on. Finally, a fair copy of the plan is prepared and perhaps displayed within the school.

Brian Spicer suggests that likely outcomes of this activity include:

1. increased factual knowledge of the Western Port region;
2. increased skill in comprehending and analysing data presented in a variety of forms;
3. increased understanding of those factors which make an area attractive for industry, residential development and have increased commercial development;
4. greater understanding of the views of conservationists, recreational users and others who see the development of a great urban complex as a threat to the resources, attractions and existing community in the area;
5. increased skill in analysing, synthesising and evaluating evidence from a variety of sources and an increased ability to reach judgements and decisions on the basis of evidence and personal considerations;
6. greater willingness to listen to the views of others, greater skill in seeking compromises, and
7. greater understanding of the complexities of regional planning and of policy formation.

There seems little doubt that these outcomes are realistic assessments of the intellectual, social and practical learning which can be achieved through involving students in geographical problem solving and decision making of the kind offered by Western Port.

Geography is rich

As I have stressed through numerous examples in this chapter, we have a tremendously solid investment in data banks. From our data banks we borrow resources to develop our students' intellectual, social and practical talents. To ignore our resources and

their potential contribution to student learning is to deprive geographical studies of their greatest strength—their close and essential connection with the world, real or imagined, public or personal.

Further reading

I suggest that, where possible and if reasonably accessible, the following journals should be consulted regularly for those in search of teaching materials and ideas similar to those included in this chapter.

I am aware that my list shows an understandable bias towards English-speaking countries and journals which represent national associations or aim at a national or international market rather than a state-wide or local audience. This is not to downgrade the latter in any sense; they are very lively and young teachers will want to link into local associations and local informal contact groups where these exist.

Bulletin for Environmental Education Published by the Town and Country Planning Association, England.

Classroom Geographer An independent journal published at Brighton Polytechnic, Brighton, England.

Geo A magazine for students published by Mary Glasgow Publications, England.

Geography Journal of the Geographical Association of England and Wales.

Geographical Education Journal of the Australian Geography Teachers Association.

Instructional Activity Series Published by the National Council for Geographic Education, USA.

Journal of Geography Journal of the National Council for Geographic Education, USA.

New Zealand Journal of Geography Journal of the New Zealand Geographical Society.

Social Education Journal of the National Council of the Social Studies, USA.

Teaching Geography Journal of the Geographical Association of England and Wales.

Teaching Geography Occasional Paper Series, Geographical Association of England and Wales.

4 Interpreting and Analysing Attitudes and Values

Attitudes mesh with thinking skills

The way values exercises are used in classrooms depends on the relative weighting and emphasis a teacher gives to (1) the knowledge and skills, (2) the attitudes and values component of learning the concepts of a discipline. In the last chapter, in *Planning for Western Port*, value dimensions were ever present and quite explicitly built into the whole activity. It was a value laden issue. The aim of this chapter is to draw attention to values and attitudes in a geographical education.

Broad aims

Acquiring knowledge and skills through geography has almost certainly been the predominant, overt and most explicit of our activities. The explicit teaching of values or teaching for values analysis, clarification, and development has received less emphasis though a concern for certain values and attitudes has traditionally been part of the vision of the concerned geography teacher. The belief that the study of things geographical would automatically contribute to developing good citizens, in Fairgrieve's (1926) sense, and that geography must by its very nature lead to international understanding may be cited as examples of geography's concern for fostering certain attitudes and values. Indeed, such claims and concerns may be seen as part of the ongoing development of a justification and rationale for geography in education grounded in what at any one time is considered to be educationally worthwhile and societally desirable (Graves, 1971).

Classroom activity

Neighbourhood preferences

The exercises based on the question, 'What things do you prefer in your neighbourhood?' (Figure 4.1) might be viewed and treated primarily as an objective assessment of the impact of a variety of land-uses in a predominantly residential area. Each of the changes—changes in appearance, traffic, people—mentioned in the checklist could be weighed up and a decision based on such a composite measuring stick reached. For example, each of the six potential changes may be considered to have an equal weighting which added together make up a composite score. Such a procedure gives an air of clinical objectivity and detachment to the enquiry.

Degrees of involvement

A feeling of detachment would be stronger if the locational alternatives were even more antiseptic and uninvolving, and were replaced by giving students a more remote adjudicating role. For instance, as town planners, they might be asked to place newspaper stands, hospitals or launderettes in a Burgess concentric ring type of town and undertake what amounts to a compatibility of land-use exercise removed from any engagement of their personal feelings in the matter.

The fact that students do have to consider the potential types of land-use change as if they were to be placed in their street and then in progressively remoter areas does probably bring about an interplay of their own attitudes and values a little more directly. Their feelings about living down the street from a hospital will interact with their more objective appraisal of such a land-use. Clearly this ever present interaction between cognitive and affective domains repudiates suggestions that values and attitudes can be kept out of the classroom.

Figure 4.1 Which things do you like best in your neighbourhood?

Would you change your neighbourhood?

Read carefully the whole list below before marking anything. Think about each item in the list as if it were something that was going to be built in the four places at the top of the columns. The aim is to sort out your likes and dislikes for such things. After you have thought about each thing, copy out the list and give a 1, 2, 3, 4 or 5 to each. A score of 1 is highly desirable, 2 is desirable, 3 is indifferent (you don't mind one way or the other), 4 is undesirable, 5 is highly undesirable.

Remember that any change affects many things. It may cause:

* changes in appearance and views
* changes in the way people travel, shop, meet each other
* changes in traffic
* changes to air, water . . .
* changes in the amount of noise
* changes in the kinds of people in the neighbourhood

	In your street	In a neighbouring street	Within another part of your neighbourhood	Within a nearby neighbourhood
1. Hospital				
2. Corner shop				
3. Bingo hall				
4. Launderette				
5. String of shops				
6. Newspaper stand				
7. Supermarket				
8. Public dump				
9. Taxi headquarters				
10. Take-away restaurant				
11. Open-street market				
12. Home for vagrants				

● Underline those things you have marked 4 or 5 and write down or discuss how you would actually behave if a proposal for building it were put forward. Decide from the clues given in A, B, C, D, E and F. A—you would move elsewhere; B—you would protest or organise a protest group; C—you would join an already existing protest group; D—you would complain but not take any action; E—you would do nothing; F—other (explain).

● Ring the things you marked 1 or 2 in the first column and work out what they have in common.

● Choose one of the things least desired for your neighbourhood by the class as a whole. Decide how the bad effects of the change could be reduced or where it might be placed outside your neighbourhood. Consider whether anyone would think you were being selfish by refusing to have it in your neighbourhood.

SOURCE: F. Slater and Moeller (1981) *Skills in Geography, Level 3*, Cassell.

Stay, move, protest

The strength of student feelings and how their behaviour might be affected is linked in the activity to the options of moving elsewhere, organising a protest group and so on. Ideas on political action may be said to have been introduced—a nod towards political literacy.

But there is no suggestion of an *overt* examination of the preferences, the attitudes and values underlying the choices. Would we, in fact, wish to probe further beneath the preference pattern? Do we see any need

to confront students with a 'why?' question and ask them to consider: 'Why? What makes you say that? Do you think that your statement holds true in other situations? What values lie beneath your choice? Are you proud of your choice? Would you stand up in public and advocate your views?' Such questions encourage students to realise and explore their opinions or attitudes so that they become more aware of the values underlying their choices. It is questions like these which would need to be asked to transform the neighbourhood land-use change exercise into a values exercise.

A moral question

Notice that a moral question related to sharing is raised, almost as an afterthought at the very end of the activity. I am now inclined to think that this question on selfishness might not be the best kind of question to help students towards a realisation of what values underlie their decisions since it is a question loaded towards one particular moral value. Probably more neutral, strictly analytical questions should be asked, at least in the first stages of values analysis and clarification. The private or public realisation that choices may reflect selfish, altruistic, generous or bigoted attitudes is appropriately reached as the culmination of such exercises. In the context of land-use change the question about selfishness needs at least to be preceded by 'Why do you think …?' questions. Acting selfishly or unselfishly and its related attitudes would come up in response to a 'why?' question. If the question were asked as part of a sequence of questions, would a teacher have to be prepared to discuss how we arrive at decisions which represent a reasonable and wise balancing of our needs against those of others? Has Western Port been wisely developed?

Values education

Students need to be made more aware of *their own* and *other* people's attitudes and values. In this chapter a number of the major strategies in values education will be presented. The implications arising from values education seem at first formidable, quite apart from immediate worries about indoctrination.

Do geography teachers now need to be well read and deeply conversant with the content and issues of morality and moral education as well as geographical education? Are there concepts within political education as a developing set of ideas of which they need to be aware? Schools have for many years assumed responsibility for the knowledge and skills areas of the curriculum but do we take greater risks and over-reach ourselves in encouraging students to examine attitudes and values? Certainly, value-laden topics and issues are more openly approached than in the quite recent past.

A new frontier

Values education seems at first to pose more questions than it supplies answers. It is the geography teacher's new frontier, to coin John Huckle's phrase (Huckle, 1976). Some will ask if indeed it is possible or justifiable to 'teach' values or to seek to influence students' attitudes and values. Can or should the teacher compete in a values forming environment in which home background, friends, TV and advertising already play such an influential role. Are not values a matter of opinion in a pluralistic society? Thoughtful answers to these questions can refute the concerns that (1) values education is indoctrinary, or (2) geography teachers can deal with their subject in a value free way or (3) the idea that values can be equated with opinions. One can argue with Huckle (1976) that an explicit knowledge of strategies for exploring and assessing values is a defence against indoctrination, that geography is not a value free subject and that there are some ultimate values and moral principles which cannot be reduced to being relatively right or wrong, good or bad, desirable or undesirable, however problematic their implementation may be.

Teaching style

However, at a pragmatic level, if geography teachers see themselves as transmission-reception teachers with a responsibility to give clear-cut answers, and people on whom students rely for absolute guidance, then they will avoid any explicit attempts to probe beneath the surface of the affective domain. If, on the other hand, they see themselves in the role of mediators, working within the interactionist model, then their responsibility becomes one of leading and facilitating discussion, developing rational argument and fostering a non-threatening and supportive environment in which students can explore attitudes and values. Geography teachers do not have to provide ultimate answers. They have to be able to challenge people to think further and encourage them to accept their conclusions as tentative and susceptible of further development and appraisal. They must acknowledge that this is going to be difficult if, in fact, they cannot hold their own views open-mindedly (Welsh, 1978).

Taking up the challenge to raise questions of value, presupposes students accept education as an activity in which they and their teachers are constructively engaged and one for which they have developed certain necessary social skills and forms of behaviour. I make the point because I think it is an even more crucial assumption in values education if such activities are to be deemed possible.

Neutralists in values education

To leave the discussion there is to admit to being one of the new neutralists—those who advocate analysing and clarifying values but not going so far as to take part in developing and forming values and acknowledging some values as absolute. There is no doubt, as already mentioned, that there are many teachers who hold the belief that values should be kept out of schools and education and that values education is an infringement of the human rights and freedom of students and parents. Another group of people believe that education of the affective domain is important and necessary. Many of these people would avoid teaching specific or ultimate values and emphasise the process of clarifying values. Raths *et al* (1966), Simon, Howe and Kirschenbaum (1972) exemplify this position. Their strategies for probing values and attitudes do not encourage or permit the teaching of values. Other writers like Metcalf (1971), who are also neutralists, emphasise rather more the role of values analysis for improving value judgements and helping students to make more rational value judgements, while Kohlberg (1975) advocates the development of moral reasoning. His view and approach is not considered neutral because the moral decisions students are asked to make involve concepts of right and wrong.

Moral reasoning

Kohlberg, like Piaget, is not without his critics, but he does, at the very least, present us with a view of moral development and a strategy for encouraging moral reasoning. It would seem that discussion of moral issues is likely to be more beneficial than detrimental if, in fact, we are willing to engage in such an enterprise. Willing or unwilling, we need to be clear about our reasons in either case and to have a broad view of our aims in relation to values education and geography. It is perhaps unfair to make a clear cut division into those who are either willing or unwilling for there are probably degrees of willingness and unwillingness to attend to values and attitudes in school.

Wilson, Williams and Sugarman (1968) identified a hierarchy of levels of endeavour in moral education. Such a hierarchy is useful as a guide to sorting out just what we are likely to be achieving and at what depth we are engaging in the interpretation and analysis of values- and attitudes-laden issues. Guided by the Wilson, Williams and Sugarman hierarchy, though not adopting it completely, I suggest that we have to consider the likely effect for values education of teaching at the following levels:

1. Factual knowledge
2. Empathy
3. Classification of values and ethical principles
4. Encouraging the exercise of moral judgement.

Factual knowledge

Students cannot be expected to reach sound value judgements without an understanding of the facts, ideas and generalisations that relate to the topic under discussion. A knowledge of factors influencing the location of agricultural enterprises, industries, high-class high-income residential areas, or shopping centres is necessary in order to reach locational judgements pertaining to such enterprises and entities. However, the effectiveness of promoting this level of understanding for values education is doubtful for the reasons I have already discussed in relation to the neighbourhood land-use change activity. But where the attitudes and values of key decision makers and pressure groups is incorporated into the data students are working with, then the exposure to the values of others enhances the quality of the learning about values and for values. Role-play games like *Planning for Western Port* which provide information on the issue, together with people's likely attitudes, and then set up decision-making activities are likely to be effective in demonstrating the interaction of fact and value. However, the aim of such work is most often directed at providing insights into decision making *rather than the valuing process as such*. Nevertheless, there are very real spin-offs in the realm of attitudes and values education. We learn that the most powerful interest groups most often win. Do we need to go further into political education to bring about an understanding of these outcomes? Do we need to go further than developing a knowledge and understanding of who holds what attitudes and what self-interests are apparent? We have long believed that we should teach for empathy, for example.

Empathy

Empathy is promoted by providing concerned insights into the situation, feelings, anxieties and intentions of other people. It involves the not inconsiderable ability to place oneself in another's shoes or life circumstances and expectations, in order to view problems, dilemmas or situations from their perspective. Can the conservationist feel him/herself into the position of the industrial entrepreneur? If so, an insight into motives and actions is achieved.

Exercising empathy requires that one can be objective enough to understand another's viewpoint without necessarily agreeing with it or accepting it as the only possible viewpoint. Games and simulations are widely used to encourage and develop empathy. The British Schools Council Liverpool team considered that simulations promoting empathy require children

> (a) to assess as accurately as possible how another person or group sees a situation, and (b) to select the elements in an often complicated situation that are related to the way another person or group *appears* to be seeing that situation.[1]

These requirements of empathetic understanding are well worth pursuing in order to understand one's own and another's attitudes and values. The second of the two points probably does not receive enough attention. Teaching for empathy under point (b) becomes a tool for examining personal values and exposing contrasts in values between individuals and groups. Teacher questioning is vital, however, and needs to play an important part in the debriefing session at the end of empathy-raising games. What attitudes and values did the different people hold? Who would most closely represent your viewpoint? What is your view? What influenced the people to hold the views they did? Why do you think the person holds those views? What arguments would you put forward to support one of the views which is contrary to your own? Would that person hold those views in all circumstances? What might cause a change of viewpoint? These questions move from closed to more open questions which might stimulate a critical awareness of attitude formation and change.

I think geography teaching has probably made its most positive and impressive impact at the level of accepting the development of empathy as a goal of learning through geography. Few writers seem to disagree that learning through geography should promote empathy, especially in working towards positive attitudes to other nationalities. Today development education is leading us to understand that positive attitudes are not enough. But there are other scales of operation besides the international which need attention and values education provides teaching strategies which may be more effective than our attempts to date.

The British Schools Council Geography 14–18 Project has published materials, for example, in the Population unit which are directed towards achieving empathy. Emphasis has already been given in Chapter 2 to the concept of cultural mismatch in views of development. Such a concept encourages an appreciation of the need to empathise. Beyond empathy there are still deeper levels to be considered.

Clarification or formulation of values and ethical principles

The clarification or formulation of values and ethical principles involves articulating and understanding the basic values which underlie and inform one's attitudes towards people, places, objects or issues. In Fenton's (1966) terminology, basic values are substantive values while Rokeach (1977) refers to them as terminal values. In clarifying or formulating their own and others' basic values, students would be consciously recognising the beliefs and ideas underpinning attitudes as these are felt or expressed in a variety of circumstances.

The bringing to consciousness and the enunciation of basic values and principles is significant for two reasons. First, it helps students to realise the processes underlying preferences and judgements. Second, the implications and consequences of holding a value or set of values can be explored in the context of a variety of problems. Perhaps students need to be given opportunities to sort out their own positions on an issue and to argue to this rather than always being asked to adopt a role.

Encouraging the exercise of moral judgement

Exercising judgement requires students to use their ethical principles to help them to understand issues and to make decisions. It is perhaps best exercised in values education through the use of moral dilemmas and an example is given later in the chapter when Kohlberg's ideas are discussed more thoroughly. If I were a conservationist in the Western Port context and yet the chances of my daughter or son obtaining a job in the area depended on expansion, what would be my stance at the public meetings?

Classroom activity

Values analysis

Let us now return to the exercise described in Figure 4.1, 'What things do you prefer in your neighbourhood?' and expand on it a little to create a values analysis activity. Let us suppose that one of the options is a home for battered wives. After asking an initial question, 'How would you feel if a home for battered wives and their children were set up in a house in your street?' as a trigger to thinking, the newspaper article set out in Figure 4.2 should be given to students to read.

A teacher-initiated discussion could clarify any points unclear to the class and help to make sure that the different viewpoints expressed have been comprehended.

The scenario now becomes a public meeting which all students as residents of the South Hill attend. A resolution is before the meeting: 'That in the interest of community standards and property values, this meeting instructs its Chairman to oppose strongly the use of a residence for battered wives and their children at a meeting with the mayor.'

The following procedure, simplified from one used so effectively by John Willmer, is suggested as appropriate to a values analysis. (For further development of this strategy see Fien and Slater, 1981.)

1. Individually students vote 'for' or 'against' the resolution and write down three reasons for their decision.
2. Those who voted alike are grouped in pairs and each pair decides on the best four reasons they have written out for their choice.
3. Pairs 'for' and 'against' join together where numbers permit and argue out and select the best reasons on each side and list the values underlying each reason.
4. The values, further clarified by class discussion, can be listed on the blackboard.
5. A second vote is now taken and students give their reasons once again in a discussion of any

changes they make to their voting and/or reasons.

Values in geography

A lesson about the location of a home for battered wives does not seem to me to be fundamentally alien to geography's concerns. It can be seen to be firmly rooted in locational analysis and scientific geography. One could speculate about why such value laden issues have not been prominent. However, curriculum units and activities for developing an awareness of values are now increasing. The diagram, Figure 4.3, based on Hurst, 1968, provides some clues when what it says is linked with the conceptual revolution in geography and what developed out of it.

Figure 4.3 The milieu in which economic activities occur

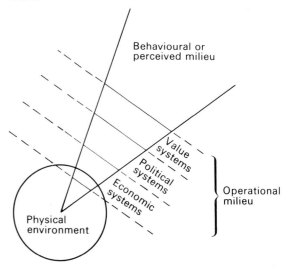

SOURCE: based on M. Eliot Hurst (1968) *A systems analytic approach to geography*, Commission on College Geography, General Series, No. 8, Association of American Geographers.

Figure 4.2 A home for battered wives

The Geraldine Mail

Haven For Battered Wives

by our Staff Reporter

The Geraldine Welfare Society's plan to set up a refuge for battered wives and their children in Test Street on the South Hill area of Geraldine is causing some local concern and unease.

Major Smith, the society's spokesman said that no final decision had been reached and that they were awaiting the report of a working party set up to investigate the feasibility and desirability of such a venture. Present indications were that such a home was badly needed. He said that his organisation had received a sympathetic hearing from the local council whose planning sub-committee would have to agree to such a use and to the external plans.

The Secretary of the Residents' Association for the South Hill, Mr Richard Rothwell, is calling a meeting of members to sound out 'public opinion' on the plan. Mr Rothwell claimed 'I have no strong views myself but I do think it is important for people living in the area to discuss and debate the proposal.' He invited all interested members of the public to a meeting in Columba Hall, on Tuesday 11 November at 7.30 p.m. A member of the council's planning sub-committee had promised to attend.

The *Mail* has conducted its own survey of local opinion and discovered that citizens have mixed feelings about the plan. 'While I sympathise with women in such a plight,' said Alicia Bottle, 'I am worried about the type of people we will be bringing into the area. We cannot be sure that some violence will not occur and our local school may have to deal with unfortunate but maladjusted children. Violence begets violence doesn't it? What sort of people will we be allowing in our area?'

Another facet of the proposal which concerned Mr Colin Dawson-Jones, a well known businessman, is the effect on property values. 'I have lived here for many years and it is a quiet, respectable neighbourhood. I am concerned that the setting up of an institution could be the beginning of a change in the nature of the whole area. I think the place for institutions is in the country where peace and quiet must have a beneficial effect on troubled people.'

Reactions were mixed among mothers of a local play group. There was some apprehension about the capacity of the group to absorb many more children, especially those from unfortunate backgrounds. 'But we do have a duty to cater for everyone's needs,' said Archdeacon Davies' wife.

The Residents' Association is hoping for a large attendance at its meeting so that the concerns people feel about the proposal can be aired.

In the first round of the new geography, there was a focus on the processes underlying locational patterns. At first, economic processes were emphasised. Investigating process soon led to an interest in decision making and the identification of the variables influencing decision making. Figure 4.3 purports to be a representation of the milieu in which economic activities occur. In its emphasis on economic activities it is very much of its time, though it also specifies in its overlapping systems and environments a hint of concerns to come. There is a values system operating to constrain and inform people in their decision taking activities.

The values system in our models, in fact, at first incorporated little apart from economic values and these created idealised people living in simplified, though conceptually useful, environments who were profit maximisers and minimum risk takers. The supremacy of economic goals and profit maximisation was so strong as to be a basic assumption in the models popularised in the 1960s. I do not denigrate such assumptions since they were necessary intellectually if some understanding of location processes were to be gained through operating simplified models of reality.

Economic man in school geography— Acme Metal

Interesting examples of this kind of thinking as it manifested itself in geographical education are to be found in many of the role plays and simulations published in the High School Geography Project resources. In the Acme Metal Company flood game structured around cost benefit ideas and calculations, most of the people are given economic goals and profit maximising/cost minimisation briefs. The following quotations support my assertion:

> ... *The Manager of Acme Metal.* His interest is to get the maximum benefit for his company at the least cost. . . .
> *The Mayor of Allentown.* He would like to get the best protection possible for the least cost to the local government. . . .
> *An Owner of Flood Plain Property.* He would like to sell or lease his flood plain land for a factory site. He hopes to be able to assure prospective buyers that there is little

danger, so he is eager that the federal government, or any government, does all it can to minimise damage. . . . *A Typical Flood Plain Homeowner.* He wants full protection for his home and his job and does not care who pays for it so long as the taxes he pays are not affected appreciably. . . .

The game of farming

The widely diffused American HSGP farming game is described in the Teachers Guide as an agricultural investment activity. The game is again interesting for the assumptions implied about maximising economic returns. No information or suggestions are built into the game to reduce the desire to make profits and such a desire might well be a realistic aim of pioneering farmers. However, even if it is not easily reflected within the game as set up, it is noteworthy that the role cards omit any reference to other values which may have affected a farmer's decision making.

A typical role card reads:

1880
ROLE CARD
Since the Civil War you have been living in the Shenandoah Valley of Virginia. You have been farming in partnership with your brother, but both families are now so large that the farm will not support you both. Your brother has given you one-half of the farm implements and horses and $1,500 cash as your share of the farm. After reading about homesteading in Settler County, Kansas, in a railroad brochure you have decided to move to Kansas.

You have had experience growing wheat, barley, rye, oats, and corn. You have also had experience raising both cattle and hogs.

The farmer's previous experience is the only variable selected to suggest influences on his initial farming behaviour. I should note that knowledge which is actually gained through playing the game of farming is made more explicit, e.g. attitudes to risks if debriefing is carefully undertaken. By the time the Schools Council Geography 14–18 Project published materials, the economic system had become a socio-economic system as Figure 2.7 in Chapter 2 shows.

In inservice work HSGP put forward a number of evaluation techniques and one in particular is significant to values education. A semantic differential to measure student changes in attitudes to farmers, as a result of playing the farming game, is set out. I see this as an example of an operational advance in the role of values and attitudes in geographical education. Where geography for that vague term 'citizenship' was once lauded, in this exercise changing/improving attitudes to some citizens, for example farmers, is made more possible through the use of a semantic differential.

Beyond economic values

People could not live by bread alone, even in models, for more than a decade however. Empirical investigations demonstrated that people may choose other than to optimise economic returns. They may decide as satisficers to limit their earning power and spend more time with their family or take longer holidays, for example. People's perceptual response to opportunities and limitations in the environment varies. To understand decision making and human actions attention has to be given to how individuals perceive their world and how they choose to act within and upon it.

Another questioning of the assumptions and methodology of theoretical geography, first apparent in David Harvey's *Social Justice and the City* (1973), examined models of urban structures and found them deficient not because they do not explain the city as perceived, but because the models are reflections of the dominant system of values in our society. The consequences of our values are undeniably manifest somewhat statically in landscapes (hedgerow enthusiasts may not agree) and more dynamically in complex human-built environment systems which are usually studied topically or thematically, for example as urban, transport, rural or regional systems.

A value free geography?

This brief sketch of the changing emphasis within explanation in geography has, I hope, established that the content and procedure of geography has never been, and cannot be, value free (Blachford, 1972). Geography is, indeed, value loaded. There are values present in what is being studied and how it is being studied.

P. M. Cowie (1974; 1978) expands the value position implicit in the decision to include geography in the curriculum. She concludes that at its core, geography teaching assumes that life is good and seeks its preservation and development in terms of earth's resources and man's potential. Environmental and social concerns within the subject reflect the value placed upon life. Within the context of geography, we are studying the spatial dimensions of not only the external world but also of people's values as they are expressed through exercising preference, choice and decision. Churches in the landscape, the distribution and organisation of new towns, the migration of peoples, all reflect people's values and are germane to a study of human geography.

Through scientific or more humanistic methodologies, certain values are again expressed. The scientific approach places a high value on the development of numeracy and analytical thinking skills, along with the use of field work for data collection and the computer for data processing. The skills and understandings promoted in humanistic geography will be different in emphasis and foster the development of feelings and conscious introspection about people and places which requires the exercise of oracy and literacy rather more than numeracy.

Behavioural and humanistic work in geography reflects a child-centred approach though the pursuit of scientific geography is by no means devoid in placing the student at centre stage as data gatherer and interpreter. To raise the concept of child-centred approaches and recall its polar opposite—traditional subject approaches—is merely to illustrate another area in which values are implicit and explicit in the exercise of geography teaching and education generally.

The development of a concern for values

Developments in university geography have created a greater awareness of value positions over the last decade and rethinking in geographical education has shown a similar concern for the subject matter of social relevance; namely, the need to educate for political awareness, to recognise the influence of political decisions on spatial patterns; and to teach values through geography, in order both to increase student self-awareness of individual values and value positions, and to consider differences in attitudes and values held by groups and individuals involved in political conflict.

School geography has probably been influenced in these interests by research geography, but as A. M. Welsh (1978) points out, one should not assume a direct causal connection. Changes in educational thinking and the various efforts to introduce more social, moral and political content into the curriculum are significant influences affecting approaches to teaching social sciences. Amongst these approaches are the strategies which distinguish the values education movement prominent in North America during the 1970s. These strategies are practical procedures through which students analyse and clarify their values. For a number of reasons then, values are obtaining a more widespread recognition and explicit place in learning through geography.

Defining values and attitudes

As such teaching and learning gains ground, there is a need to be clearer about the definitions ascribed to such concepts as values, beliefs, attitudes, opinions and preferences. Along with feelings and emotions, such concepts are part of the emotional make-up of one's identity, and it seems clear that all are learned predispositions. We are not born with a set of values and attitudes, they are socially acquired. What then distinguishes values and attitudes?

Consider that you have been asked to rank the concepts on scales marked from high to low as in Figure 4.4. The scales measure the level of stability, personal commitment and predisposition to action or involvement associated with each concept. It is likely that on each scale values would be ranked higher than attitudes.

Values are the more stable and enduring of the two concepts, most probably because they are initially taught and learned in isolation from other values and in an absolute all or nothing manner. Honesty, for example, is always held to be desirable in all circumstances, it is not just sometimes desirable. Our attitudes towards something can change more readily than our fundamental values.

The degree of personal commitment with which we hold to values and attitudes varies and there is a relative quality present. Situations will arise in which several values may be in competition. One value will have to be weighted against another and the hier-

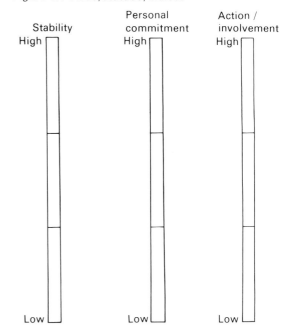

Figure 4.4 Value/attitude/indices

SOURCE: based on an idea by J. Fien.

archy into which we organise our values may be rearranged from time to time.

There are values which are at the very centre of what we consider to be important in human existence, others lie more peripherally and receive less personal commitment. At the level of action and involvement it is difficult to be certain, as seems logically likely, that there is a strong relationship between taking action and the most stable of our values. Perhaps we want to act when our most strongly held values are brought into play by a situation or circumstance but understandably the relationship is not that simple or direct. The behavioural or action component worries us most as teachers. We step back from possibly affecting people's behaviour outside the classroom though we take responsibility for moulding social and cognitive behaviours within the classroom. Fenton's category of procedural values is a useful concept in this context (Fenton, 1966). We are prepared to encourage, for example, a respect for evidence and co-operation through group work.

The two concepts have, then, a number of things in common. How do they differ? First in number. We have few values, but many attitudes. Rokeach (1973) defines a value as an enduring belief that a specific mode of conduct or end state of existence is personally or socially preferable to an opposite mode of conduct or end state of existence. Instrumental and terminal are the technical terms he uses to refer to modes of conduct on the one hand and end states of existence on the other. We may ultimately value a comfortable prosperous life and work with ambition and diligence to achieve it. We have a conception of and a preference for the 'desirable'.

What then are attitudes? Attitudes are defined as packages of beliefs which influence us in decisions. Attitudes are those beliefs which, when focused on a specific object or situation, predispose one to act in a preferential manner. It is the idea of revealed preferences in relation to an *object* or *situation* which distinguishes attitudes from values. The first are more specific, the second more general in applicability. Attitudes are value expressive.

Classroom activity
People or plants?

The following script written by R. G. Richmond as part of an assessment unit on the effects of erosion by trampling on Hampstead Heath demonstrate, like many other such roles, the activation and exposure of attitudes. Underneath attitudes lie less clearly discernible values. This assessment would follow (1) work on erosion generally, (2) observation and measurement of erosion on the Heath (data being subject to a chi-square test) and (3) interviews with park personnel and people using the Heath. It is intended to make students aware of the attitudes and values of others, to cause them to consider their own and to set out a reasoned argument for their proposed management strategy. Richmond's scenario goes like this:

You are the Park Manager in charge of Hampstead Heath, and at the Policy Planning Committee Meeting following the publication of research findings, you are the Chairman. You will be required to reach a decision on future policy towards erosion caused by trampling based on points made by different interest groups at the meeting.
Around the table are:
(a) representatives of local residents' committees variously concerned with
(i) conservation on the Upper Heath, and
(ii) recreation amenities.

(b) a statistician (purely an expert on data interpretation; no knowledge of the area or issues involved).
(c) a representative of 'big business' (has no concern with the interests of the Heath but would like use of a prime site adjacent to Hampstead Village and actually on the Heath to build a new hypermarket).
(d) YOU, THE PARK MANAGER (you are required to sum up the views expressed in the meeting. NB. The Heath is expressly for the use of the people of *London*, i.e. not *just* the local area).
In addition members of the public attend the meeting. Speakers from the floor include:
(a) Mr George Smith, who lives with his wife and baby in a privately rented basement bedsit in the vicinity of the Heath (speaks for unrepresented working classes).
(b) Extreme recreationalist: feels more recreation facilities should be available on the Heath (hypermarket boss may offer financial assistance).

Excerpts from their representations
1) *Miss Prudence Hetherington-Chumley*—Highgate Society, on conservation of Upper Heath ... I think it is essential to maintain the rich natural environment of the Upper Heath with its fine balance of wild grassland, woodland, gorse and heather. Certain users of the Heath are currently wrecking the vegetation and laying the ground bare to the elements. I propose that large areas of the Upper Heath be closed to the general public in rotation to allow regeneration of plant species; the public should be kept to proper footpaths or kept from an area altogether. However, access should be available to ecologists and other special groups, perhaps monitored by a pass system ...
2) *Miles I Chase*—Heath and Old Hampstead Society—on recreation ... I totally disagree with Miss Chumley. The Heath is an amenity for the people of the area, and no section should be closed to the public at any time. Suggestions on use might be made, e.g. signposted routes for joggers, cyclists etc. or marked-out football and rugby pitches, but no parts should be inaccessible to any user. The Heath is here to be used and enjoyed, not be protected from all human activity. So no tarmac, no fences, and certainly no no-go zones.
3) *Gordon Bennett* (Statistician) ... The results of the statistical analysis show quite clearly a relationship between trampling of one kind or another and erosion. Of course, one doesn't wish to imply causality, and further statistical testing may be required, but it does seem that if we're to stop the erosion we must stop the trampling, and the only way

to stop trampling is to stop people using those areas where erosion has been shown to occur.

4) *Antonio Tusko-Soansbury* (Director, and developments specialist) ... Well, we at Soansbury's are very sympathetic to the needs of the Heath, both in terms of amenities and conservation—but we're also aware that the costs of employing sufficient human and technical resources to achieve the ideal balance are too great for the G.L.C. to be the sole provider. My company would be prepared to offer $1.5m of immediate assistance and a further $300,000/year towards achieving the goal in exchange for about 20 acres of the Heath immediately adjacent to Hampstead village. The plan is to develop a new hypermarket here; it would *of course* be in harmony with the existing environment, with landscaping etc.

5) *George Smith* (local resident) ... I think you're all out of touch. You talk of jogging versus conservation as if no other conflict arises, whilst people like my family live in a damp basement in an overcrowded house just down the road. What this area clearly needs is more houses; they've got to come before parks and things. But the people who make decisions or influence them are the middle classes, already comfortably housed, so the real conflicts are brushed under the carpet.

6) *Christos Christodoulos* (local resident) ... George has got a point, but you can't just have houses. People must have somewhere else to go. I think there really aren't enough amenities round here and they should use the Heath to develop more; proper running tracks, a large swimming pool, indoor sports hall, etc. The businessman's money should be put towards that. Give people something to do and they probably won't trample all over the rest of the Heath.

On the basis of these representations, the statistics presented at the meeting (plus your reservations) and your role as the manager of an area of parkland expressly for the use and amenity of the people of London, write a reasoned argument summarising and criticising the main points presented above, and leading to a decision regarding treatment or otherwise of those areas exposed to erosion by trampling on the Upper Heath (see also *Classroom Geographer*, May 1981).

Promoting value awareness

This activity is an exercise in values education. It seeks to make students become more aware of their own values through the presentation of attitudes likely to be held by others.

Figure 4.5 A scheme for values/attitudes analysis

> 1. *Analysis of the situation.* What problem faces the person or group? What values are implied?
>
> 2. *Analysis of value options.* What alternatives are perceived? What alternatives are possible? Are there possible alternatives not perceived?
>
> 3. *Rationale for the decision.* Why was that alternative chosen?
>
> 4. *Consequences of the decision.* What are the consequences of the decision? Can it be verified that the consequences result from that decision/value?
>
> 5. *Evaluation of the decision.* Which alternative would you have chosen? Why? If your choice matches the decision made, are your reasons the same? If your choice is different, explain why.
>
> 6. *Justification of the values.* What justification can you give for the criteria used? Do you use this value as a criterion consistently? Does it fit in with your other values? How?

After a decision has been reached in a values attitude conflict exercise, the scheme outlined in Figure 4.5 could be used as a means of a values/attitudes analysis. We need to be careful to develop questioning which does more than elicit attitudes (to objects or situations) and digs deeper into values. It would seem that a number of exercises used at present are weighted heavily towards the attitude end of the scale and demand pragmatic decisions based on compromise which could lead to values confusion rather than values clarification and development.

Strategies in values education

Quite definite schools of thought and procedure, briefly referred to earlier in this chapter, have been developed by various educators as a means of making values education a practical undertaking. Thus far I have presented exercises involving an examination of values and attitudes. I shall now explain the different schools of thought and highlight their procedures through examples. The values clarification approach developed by Raths and colleagues is fully set out in *Values and Teaching* (1966). The clarification school holds that values must be subjected to three processes

Figure 4.6 Valuing processes

PRIZING one's beliefs and behaviours
1. prizing and cherishing
2. publicly affirming, when appropriate

CHOOSING one's beliefs and behaviours
3. choosing from alternatives
4. choosing after consideration of consequences
5. choosing freely

ACTING on one's beliefs
6. acting
7. acting with a pattern, consistency and repetition

SOURCE: based on Raths, L. *et al* (1966) *Values and Teaching*, Merrill.

set out in Figure 4.6. Seven criteria define the processes. To use this strategy in relation to the siting of a home for battered wives the following questions matched to each criterion could be used. Obviously in the classroom there is an element of role play in the prizing and acting levels for clearly students may or may not be prepared to hold such values outside the classroom or affirm or act upon them in a real life situation.

Classroom activity

1. If you were the mayor of Geraldine Borough Council, what would you do to resolve the question of the setting up of the home?
2. What do you think will happen if the plan you put forward is put into action?
3. What are the advantages and disadvantages of the position you have taken?
4. Why did you reject alternative ways of handling the issue?
5. You have to present your solution to the issue to a meeting of the Council. Write out a short statement of exactly what you are going to say.
6. Pretend you are a relative of a battered wife. What would you do to get the Council to accept your suggestion?
7. The problem of allowing a home for battered wives may occur again in the Borough. What policy or plan do you think the Council should set up to handle it in the future?

Preparation time

Raths *et al* (1966) in their classroom procedures of values clarification suggest that students need time to think about their feelings and attitudes before a discussion. They suggest that students be provided with a list of questions to provoke thought and reaction prior to a class discussion and that written answers should be handed in. The significance of the written answer lies in the fact that it gives pupils the chance to think out their ideas thoroughly before discussion, and class dialogue may then be more purposeful and thorough. There are other numerous procedures which fall into the category of values clarification. These are outlined in Simon *et al* (1973), but most require considerable modification for use in discussing things geographical.

Values analysis

Another method for probing values and attitudes, the values analysis approach, was first fully described in the *1971 Yearbook of the National Council for Social Studies* edited by Lawrence Metcalf. Exercises are designed to have students arrive at value judgements. The way in which the earlier exercise on the home for battered wives is structured represents a values analysis strategy.

Classroom activity

The Coin Street controversy

The present Coin Street controversy on London's South Bank could be subjected to a slightly different form of values analysis where the essential exercise of making a value judgement centres on the land-use conflicts in the inner city. The extract from the *T.C.P.A. Journal*, Figure 4.7, gives the flavour of the planning conflict in the area. A number of pressure groups including the Coin Street Action Group, the Lambeth Council, the G.L.C., private developers, the Town and Country Planning Association, are all involved in the issue of whether 16 acres of inner city London, located within a few minutes of the Waterloo tube, should be used for homes or offices. Students could be told to assume the role of the Secretary of State for the Environment. He has to decide whether planning permission is to be given to Commercial Property Ltd and the Heron Corporation, or whether the Waterloo District Plan of 1978 is to be upheld. A list of *statements for* planning permission and *statements against* planning permission should then be drawn up by the students in their capacity as Secretary of State to resemble the following list:

Statements listed by the Secretary of State for the Environment

Statements for	Statements against
1. Proximity to Waterloo Station gives easy access to the potential office workers.	1. There are already too many office blocks in the city.
2. Commercial development may encourage a Pompidou effect.	2. There is a shortage of housing in the Waterloo area.
3. What is now housing does not have to remain so.	3. Parts of the inner city need to be retained for residential use.
4. Property development means private investment and a saving on public funds.	4. The Waterloo District Plan was agreed as a result of public participation and a thorough hearing— and should not be overturned.
5. Rates realised on commercial properties provide greater revenues for Local Council.	5. Housing proposals retain more open space in the area.
6. Office blocks could be in harmony aesthetically with the Royal Festival Hall and National Theatre complexes.	6. Office development will increase congestion in the area.

Figure 4.7 Waterloo belongs to ...?

After individual lists of statements have been prepared a discussion to clarify their meaning and assess their relative importance needs to take place. Judgements should then be made and the exercise concluded by the class drawing up a common list of observations and recommendations on the basis that the planning applications have been granted. Possible observations and recommendations may read:

Possible observations

1. It will be difficult to uphold the present development plan system.
2. Keeping a balance between commercial and residential interests in inner cities is very difficult when decisions affect small parts of the city in isolation from any reference to overall trends and development.
3. Decisions are often made on the basis of political pressures.
4. It is extremely difficult to settle issues like this in a way that satisfies minority and majority rights.

Possible recommendations

1. An independent commission, headed by a prominent urban geographer, needs to be set up to review inner city redevelopment policies.
2. Limits should be set on the amount of money which can be spent by interested parties in such hearings.
3. Political studies should be included in a core curriculum.

Fun and Survival in the Inner City

There is an undiscovered law of economics which can be expressed as follows: Finance is available for developments in the inner city in an inverse proportion to their usefulness to the people who live there. Just as any Experimental Proto-type Community of Tomorrow that can attract half-a-billion dollars is likely to involve Mickey Mouse and roller-coasters rather than the more mundane problems of people living in groups, so the dreams of property developers usually have more to do with medieval banquets and what can be got out of the inner city, than with ways of putting resources into inner areas and helping people who are suffering from the effects of economic change. Though both go under the name of investment, the two are not the same.

That is not to say that the functions of a part of a city should not change. It may well be appropriate that a particular area that once supported housing or industry should come to fulfil city centre functions, or vice versa. But who decides what is appropriate? That question is often complicated by the present system of land valuation (which effectively limits the uses to which particular inner city sites can be put) and by the political power of potential private investment. When the Waterloo District Plan was adopted in 1978, the residents of the area around Coin Street on London's South Bank saw it as confirmation that it was they who had been given the chance to decide. The plan had been produced by Lambeth Council after five years of consultation and participation, and the residents felt that it reflected their preferences. Since then, events have shaken their confidence in the plan ning system as they understood it. Commercial Property Ltd and the Heron Corporation have submitted planning applications for developing the Coin Street site; but whereas the district plan earmarked the site mainly for family low-rise houses and public open space rather than offices, the developers' proposals are largely for offices, showrooms and business suites. As this would be a major departure from the statutory plan, the Secretary of State for the Environment has the opportunity to decide for himself whether planning permission should be granted; and after a public inquiry which starts in May, that is what Peter Shore will be doing. He will not find the decision easy. On one hand, the development proposal offers the prospect of private investment in a part of London where the government is most anxious to attract it. On the other hand, this is an important test for the new development plan system. The Waterloo District Plan was one of the very first to be adopted under the new system, and if the Secretary of State does not uphold it, the system itself will lose credibility. It is this that has persuaded the TCPA to make representations to the public inquiry, in support of those who will argue that planning permission for the commercial development proposals should be refused.

After five years of hard work trying to get their vision for the Waterloo area reflected in the district plan, the residents now have to fight their case in a very different arena. The public inquiry will be considering development proposals submitted by local residents, and others by Lambeth Council, as well as those of the developers. Margaret Mellor, Secretary of the Waterloo Community Development Group, highlights a major problem; "Commercial Property Ltd, the Heron Corporation and Lambeth Council have each appointed QCs and supporting teams to prepare and present their cases at the public inquiry. The Greater London Council has put its planning, valuation and legal staff at the disposal of the private developers. In contrast, the local community groups rely solely on volunteers. Hiring counsel on a similar basis to the other three planning applicants would cost in the region of £25,000. It seems clear that the public inquiry system is weighted very heavily in favour of commercial developers and against the general public." The TCPA agrees, and has recently called on the government to provide funds for objectors at certain public inquiries as an experiment. This would be a first step in looking for ways of removing a major flaw in the public inquiry system.

Like many other residents of inner cities, the people of the Waterloo area are having to fight hard for the survival of their community in the face of the powerful forces of economic change. They look to the planning system to provide a framework for resolving the inevitable conflicts in this process, and have often been disappointed. In 1977 they heard the government's promise of a "long-term commitment" to the inner cities, and the call in the report of the three inner area studies for a "total approach"; two years later, they find few signs of either. As the TCPA says in its policy statement on The Inner Cities, there are certain matters in which the government's present policies seem to offer little prospect of improvement; namely "the need for a planning framework, public participation, new institutions to tackle problems of an unaccustomed type and magnitude, the valuation and disposal of land, employment policy, resources, infrastructure and communications". Apart from that, Mr Shore, everything is fine.

SOURCE: Cowan, R. (1979) *Town and Country Planning,* Journal of the Town and Country Planning Association, Vol. 48, No. 1.

Any number of public issues of a controversial nature are suited to values analysis including, in England, the third London airport, Docklands redevelopment, the route of the M40, mining in the Vale of Belvoir, the siting of oil refineries and nuclear power stations, as well as the kinds of conflicts present in National Parks. Similar land-use conflicts occur in many contexts throughout the world and background material is provided in Figure 4.8 for a planning dilemma specific to Dunedin, New Zealand, but widely experienced in many parts of the western world. This can be analysed in either the same way as the battered wives activity or the Coin Street controversy. There are, as mentioned in Chapter 2 in the town square exercise, a number of clearly defined interest groups—citizens, retailers, planners and so on, and public meetings or executive decisions can be employed as mechanisms to resolve conflicts and clarify value positions.

I think it is important to include this antipodean resource as an antidote to the study of New Zealand as a sheep-filled Canterbury Plains. Furthermore, it alerts geography teachers outside that country that other teaching topics are applicable and just as close if not closer to the reality of living there where more than seventy per cent of the people are urban dwellers. Similarly, *Planning for Western Port* (Spicer, 1977) may encourage non-Australian teachers to realise that deserts, minerals and the outback are not the only topics relevant to the study of that sub-continent.

Figure 4.8 Urban planning example

Planning Comment: Why not a Mall at last for Dunedin

By Richard Welch, Department of Geography, Otago University

Do you remember the Dunedin Chamber of Commerce centennial publication of 1961? The centre-fold contained a challenge from the Otago branch of the Institute of Architects to the City of Dunedin to produce new and exciting ideas for the development of the Octagon. Do you remember 'Our Kind of City' (1968), 'People and Traffic' (1971) and 'Downtown Dunedin' (1972)? And more recently, 'Plantalk 76'?

One thing these publications had in common was their call for some degree of separation of traffic from people in the central city area. Proposals ranged from the complete removal of traffic to form a pedestrians-only shopping precinct, to a rather more civilised sharing of the main shopping street by pedestrians and essential motor traffic.

Little if anything has resulted from these proposals put forward by architects and city planners and supported by Dunedin citizens. The reasons for this lack of activity are not difficult to pin-point. Retailers established in George and upper Princes Streets have been understandably unsure about, and even hostile towards, major changes to the structure of the central shopping area—at least to those proposals put forward by the city planners.

The City Council would appear to have been unable to see beyond the immediate costs of proposed changes—in particular the numbers of street parking places to be lost in implementing each new proposal. Indeed, the question of providing car-parking facilities in the central shopping area would seem to have been the principal subject of discussion between the two groups.

The buoyant economic circumstances of only two or three years ago, however, are now past....

The distinct possibility has arrived that the motorcar will become a very expensive mode of transport within ten years. If this proves to be the case, individual mobility will be reduced, as will the need for ever-increasing car parking facilities in the central city area.

As the demand for car parking space either reduces or is held constant, however, there is likely to be a concurrent expansion in demand for an efficient public transport system which will take people to the heart of the shopping zone rather than to its periphery....

POSSIBLE STRATEGY

Let us examine a possible strategy for the development of Dunedin's shopping centre, which takes into account the principal fears of retailers, the stated desires of the people of Dunedin and the uncertainty over our future individual mobility.

The requirements of such a plan would appear to be as follows: (1) It should provide some separation of people from traffic, thus permitting use of enlarged pedestrian areas; (2) use of public transport should be encouraged with all buses passing through or around the shopping zone rather than terminating at either end of the zone; (3) through

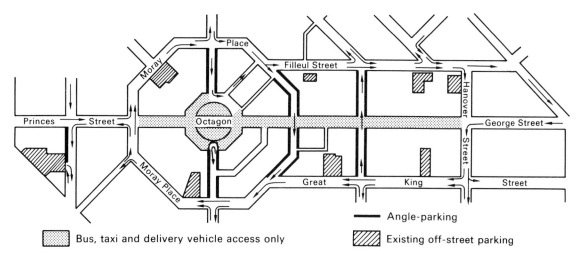

Bus, taxi and delivery vehicle access only

Angle-parking

Existing off-street parking

traffic should be able to flow freely AROUND the shopping zone and not THROUGH the shopping zone; (4) delivery and emergency service access must be maintained to all retail outlets; (5) access should be possible to all existing off-street parking areas; (6) the number of existing street parking places in the central city should not be reduced; (7) the overall cost should be kept to a minimum.

Consider a one-way system of traffic flow in a clock-wise direction around Moray Place, Filleul Street, Hanover Street and Great King Street as shown in the diagram. Private vehicles would be denied entry into George Street or the Octagon, although Princes Street and George Street would continue to be used by buses, taxis and delivery vehicles. Moray Place North and St Andrew Street would become one-way parking precincts with angle parking on both sides of the road. Stuart Street, west of Octagon, would also be a one-way parking precinct with an exit via Harrop Street, and Lower Stuart Street would become an angle parking cul-de-sac.

IMPLICATIONS

What are the implications of this briefly outlined proposal? (1) Buses and taxis can use the existing street system, bringing people directly into the shopping zone, and trolley buses

can continue to operate; (2) emergency and delivery vehicles can continue to use the existing street system—costly rear access to retail outlets is not necessary; (3) existing abnormal functions such as Festival Week processions can continue to operate; (4) access to all existing and planned off-street parking areas is maintained from the one-way system; (5) existing street parking spaces within the central shopping zone are not reduced, merely redistributed, (there are 250 existing street parking places within the shopping zone, but with angle parking in Moray Place North and St Andrew Street covering existing parking areas, and the considerable areas at present taken up by approaches to traffic signals, at least 250 street parking places can be allotted under the proposed scheme— there is, therefore, no extra cost to be incurred by providing new parking areas, the principal stated concern of the City Council); (6) perhaps most important, Dunedin could have a shopping area which does not exclude motor traffic, but which places it firmly in its (second) place.

It is not the purpose of this article to suggest the colour of paving stones in a new George Street, or to detail the types of vegetation most suited to shopping areas. Exciting possibilities have already been dis-

cussed in the 'Downtown Dunedin' report.

Of greater concern has been to present a simple, workable proposition; one which would not become redundant should the petrol-driven private motor vehicle become a luxury few of us can afford.

The question is not simply 'mall or no mall', 'traffic or no traffic', 'bustle or emptiness'; we can have all of these in our city centre, but in a combination which could be the envy of other New Zealand cities.

Within six months, with some minor alterations to existing road signs and markings, Dunedin could have a central shopping zone in which traffic movement is considerably curtailed, but which provides for all the stated requirements of existing business.

The complete re-humanising of George Street and the Octagon, with the construction of wider pedestrian areas and the introduction of some vegetation does not have to take place immediately, but can be undertaken as and when the City budget permits. The present Government subsidy to local bodies employing otherwise unemployed manpower, however, would suggest that this is a good time to undertake even this labour-intensive activity.

The challenge is there. 'Plan-talk '76' showed that the people of Dunedin want a shopping mall.

SOURCE: *Otago Daily Times*, 15 June 1978.

Moral reasoning

The moral reasoning strategy is based on the premise that individuals move through a number of stages of differing sophistication in moral reasoning, and practice in moral reasoning will lift people to higher levels. Kohlberg's research has designated six stages of moral reasoning as follows:

Stage 1   The punishment and obedience orientation. The physical consequences of an action determine whether it is good or bad.

Stage 2   The instrumental-relativist orientation. Right action consists of that which instrumentally satisfies one's own needs and occasionally the needs of others.

Stage 3   Good boy—nice girl orientation. Good behaviour is that which pleases or helps others and is approved by them.

Stage 4   Law and order orientation. There is orientation toward authority, fixed rules and the maintenance of the social order.

Stage 5   The social contract, legalistic orientation. Right action tends to be defined in terms of general individual rights and standards. There is a clear awareness of the relativism of personal values and opinions and a corresponding emphasis on procedural rules for reaching consensus.

Stage 6   The universal-ethical-principle orientation. Right is defined by the decision of conscience in accord with self-chosen ethical principles.

Kohlberg's work suggests that the stages of moral reasoning may be moved forward by providing opportunities for moral reasoning. He developed the moral dilemma, a story in which there is built in conflict over what is right or wrong. The story described in Figure 4.9 is an example of a moral dilemma set in Japanese late twentieth century society. After students have taken a position, class discussion should be directed towards keeping the discussion to the point and asking questions to introduce arguments appropriate to higher stages than students seem to be at. It should be noted that it is the type of argument and not the content or position that indicates the stage of moral reasoning.

Figure 4.9 A moral dilemma

SUMIKO SEKO'S DECISION

By the late twentieth century, the Japanese government determined that to maintain its contacts with other areas of the world a large new international airport was needed. Japan's successful growth, the government reasoned, depended on its ability to do business with companies located in other nations. Japan's foreign business dealings were essential to maintain strong economic growth in Japan.

The government wanted to build the new airport between Nagoya and Osaka. It established a three-member committee to make the final decision about the site selection. One of the three members of the committee was Sumiko Seko.

The hearings lasted one week. During that time, farmers from the area pleaded with the committee not to build the airport. The land had been owned by these families for centuries. They and their ancestors long before them had lived there and worked the land. To pave it over with runways would be to pave out their past. They felt Japan's progress had been proven. Should the country's traditional values, they asked, be destroyed?

Businessmen also testified to the committee. They said that Japan's very existence in the world today depended on its ability to send Japanese businessmen to other countries, and to receive businessmen from these places. The country's current airports could no longer sustain the load placed on them. Any delay by the committee in deciding on this new site could cause damage to the economy. Japan was in the twentieth century now, they said. Its people, including its farmers, would be best served by this new airport.

After the hearings the committee met in private to make its decision. One member voted for the site and another voted against it. Sumiko Seko could not decide as easily. She knew that if she voted not to approve this site for the airport, the farmers' land would be preserved, and the grounds of their ancestors would not be disturbed. To approve this site for the new airport would allow the country to continue its contacts with the rest of the world—something both the businessmen and the government felt was essential. For Sumiko Seko, both the land and the international contacts were important.

Should Sumiko Seko vote to approve this site?

SOURCE: Backler, A. and Lazarus, S. (1980) *World Geography*, Science Research Associates.

Types of questions

In fact, Allen (n.d.) makes a useful distinction between the various stages of moral development and the kinds of value-laden statements teachers are likely to hear. He suggests four levels of value statements each requiring different clarifying responses and questions from the teacher.

Level I Expressive-evocative statements are immediate responses to the issue at the level of a 'gut reaction'. They will generally be non-reflective expressions of feelings and attitudes. To make feelings conscious and elicit reasons is the task.

Level II Evaluative-prescriptive statements include judgements based on criteria of goodness/badness, desirability/undesirability. Students should be asked to justify their statements.

Level III Ethical statements give reasons for evaluative or prescriptive judgements and the reasoning can be clarified and expanded.

Level IV 'Life goal statements' may be forthcoming. 'I believe in equality . . . freedom . . . peace.' Such statements should again be questioned and students encouraged to explain their reasons.[2]

Teaching public issues

Another notable example of a teaching strategy, which the authors claim assists in the development of moral judgement, deliberately seeks to create cognitive dissonance or mismatch. Simon and Wright (1974) in the context of teaching public issues in the American high school have developed a useful outline for an enquiry process which is another version of values analysis. Their approach focuses on questions of desirability and feasibility—'Is it right?' and 'Is it practical?' Students are asked to decide on their position *before* collecting, classifying, analysing and evaluating data. By this means, it is likely that the solution the students come to will be different from their original judgement and create cognitive conflict.

Classroom activity

A fourth terminal at Heathrow Airport?

A. M. Welsh (1978) has used a Simon and Wright procedure to examine the proposed building of a fourth terminal at Heathrow Airport. Resources are referred to but not all are included here. The suggested procedure is:

1. Identify the problem, phrasing it in the form of a policy ('should') question. 'Should the fourth airport terminal be built?' Study the airport map and the British Airport pamphlet on the need for a new terminal.

2. Formulate an hypothesis on the desirability and feasibility of the policy contained in the question, 'Is it desirable (i.e. right) to build the terminal? Is it feasible (i.e. a practical proposition)?' Students record their opinion/hypothesis before making a systematic study of the evidence. They begin in this way to feel what the issue is about and to define their own attitudes towards it e.g. Will it improve human welfare? Will it be environmentally harmful?

3. Collect a representative sample of data and classify the evidence into:
   (a) The desirability of the new terminal (in the national interest, to protect employment, for passenger comfort, to retain tourist traffic etc.).
   (b) The undesirability of the new terminal (e.g. pollution, pressure on the green belt, pressure on public services and housing).
   (c) The feasibility of the new terminal (land is available within the airport boundary, the A30 has been improved, a new motorway which will take extra traffic is being built).
   (d) The unfeasibility of building the new terminal (only a temporary solution, not part of a long term airport plan, attitudes of different groups, alternative site at Perryoaks).

4. Evaluation of the data and hypothesis.

5. Preparation of a course of action towards the problem after examining the desirability and feasibility of taking overt group action on the problem. The majority of teachers are not likely to wish to proceed to this point. Instead students could examine the course of action already undertaken by the Secretary of State for the

Environment. Students could consider why the Public Inquiry of 1978 took place and what procedures were followed.

6. Acting on the problem, evaluating the action. Again the decision to hold an enquiry and decisions reached by the enquiry could be evaluated. What are society's dominant values?

I suggest that this model of desirability/feasibility analysis should be applied to the article, 'Aramoana Smelter Issues Reviewed' presented in Figure 4.10. It is, by chance, written from such perspectives and its clarity of style and argument make it a very suitable classroom resource though naturally for use in junior classes it would need to be simplified and shortened.

Figure 4.10 An economic/environment example

Aramoana Smelter Issues Reviewed

By Professor R. G. Lister, Professor of Geography, University of Otago; Chairman Technical Advisory Committee, Dunedin Metropolitan Regional Planning Authority

There are several critically important questions to be answered in relation to any major industry, particularly aluminium smelting, at Aramoana and aspects of each have been widely discussed during the past weeks.

From the community's point of view it is essential that we are as well informed as possible over the issues being raised and that we understand the alternatives fully. These issues are:

1. Is it worth it for the national economy?
2. What does it offer Otago as a major industry?
3. Is there enough electricity?

'No Need To Discharge Water Into Harbour'

4. Is Aramoana a desirable site?
5. Can environmental problems be resolved?
   1. **Is it worth it for the national economy?**
   Since this matter was first mooted some months ago, the waters have been ruffled by Professor von Moeseke's economic appraisal of aluminium smelting anywhere in New Zealand and it is now vital for Treasury to rethink their approach in the light of his views. . . .
   2. **What does it offer Otago as a major industry?**
   Professor van Moeseke's report (his page 10) states in referring to

regional aspects: 'The smelter not only pays wages but also purchases supplies in New Zealand, total payments for both items amounting to at least $25 million or at least $30 million in 1980 dollars at a very rough estimate . . . it is quite true that the smelter would pump at least $30 million into the economy in payment of domestic, mostly local, labour and other inputs . . . neither the regional planner nor the local authorities could, in all fairness, be faulted for trying to attract this industry since in this way the above income would be generated in, say, Otago rather than somewhere else.'

Over and above this regional input, the Government is insisting on downstream industries which would bring substantially more employment and income to the region . . .

'Workforce Could Live Elsewhere Than Aramoana'

Dunedin has a number of engineering industries that would no doubt be ready to expand the range of their activities to include aluminium possibilities. . . .

3. **Is there enough electricity?**
It will be necessary to find some 4,000 megawatts over and above New Zealand's present total consumption of about 15,000 megawatts if there is to be both a third potline at Tiwai Point and another smelter. Evidence at present shows that this is possible, though there appears to be doubt about supplying a smelter which would produce twice Tiwai Point's output—which is one of the options being considered. The authorities are re-examining the available electricity during the 1980s to cover contingencies which would be likely to occur in dry years. A clear answer has yet to be given.

An important thing to remember is that by the end of the 1980s, Otago will be requiring substantial energy resources for the development of its forest-based industries to use the resources that are being built up each year by our planting programmes.

Both aluminium and forest industries would continue to require power resources well beyond the end of this century and what we need therefore is long term planning in terms of several decades ahead.

'Picture Different From Situation Some Years Ago'

I see no reason why we should not anticipate the construction of several low dams on the Clutha as a sustained programme, each bringing in an increment of power to meet future requirements....

4. Is Aramoana a desirable site?

This issue is the most important from the view point of Dunedin people and there are many considerations under this heading which must be examined carefully.

Aramoana is clearly desirable from three points of view. Firstly, it is a first-rate industrial site in terms of its location beside deepwater with no surge problems, close to a container port. It is rare to find such advantages occurring together anywhere in New Zealand for an industry needing import/export facilities. It is possible to bring energy to the site without long transmission lines that add to costs and lead to 10 per cent or more of lost energy. With urban facilities at no great distance in Dunedin City, this adds advantages for housing and servicing.

Secondly, it is a first class biological site in respect of both vegetation and wildlife.... It is unrivalled in Otago as an example of virtually undisturbed estuarine life and it is important for the early stages of the life-cycle of at least some of the fish species found off our coasts.

Thirdly, it is rated first quality for scenic and recreation purposes as a coastal reserve in the Lands Department's Coastal Report in which it is ranked as of national importance....

The issue then becomes one of compatibility between these three different land uses. Can they be reconciled?

In 1974 when the earlier attempt was made to establish a smelter at Aramoana the Regional Planning Authority undertook a detailed investigation in respect of each of these alternatives. There was time to do the job properly and postgrad-uate scientists were employed to investigate the geology of the site, the vegetation types and the biology of the area.

This investigation was financed by the Otago Harbour Board and was undertaken as a completely independent study by the Technical Advisory and Consulting Service of the University. Its findings revealed important conclusions, revealing three component areas at Aramoana.

The landward side of Aramoana below the steep hill slope was found to possess characteristics sharply different from the remainder. This strip proved to be 80 hectares in area (200 acres) of dry firm quartz sands of little agricultural value and with low biological interest....

The seaward front facing north comprises recent wind-blown sand dunes largely fixed by marram grass and lupins over the past century and it has been built out as a result of the mole constructed to protect the deep water harbour entrance. This is the popular recreational area....

The tidal flats are the rich areas biologically, flushed by the daily tides, used by the variety of wildlife and exhibiting scientific and educational value.

'Question Of Dangerous Fumes Can Be Avoided'

EFFLUENT DISCHARGE

In any consideration of industrial sites, wind systems are important to understand in relation to possible effluent discharges.

In 1974–5 there was time to investigate the incidence of dominant winds. It was found that for some 10 per cent of the year there are calms at Aramoana while for 38 per cent of the time north-easterly winds predominate and are funnelled between the hills of Otago Harbour towards Port Chalmers, Ravensbourne and Dunedin.

It is clear that any industry producing dangerous fumes would be unacceptable under these conditions. This ruled out any industry with a chimney such as that at Tiwai Point because it would not be able to disperse fumes satisfactorily.

Discussion at that time, in 1974, hinged on the possibility of a pipeline out to sea taking the fumes beyond the local circulation of Blue-skin Bay waters. This would have been expensive but was regarded by the Regional Planning Authority as critically important.

'Aramoana Site Desirable For Three Reasons'

Since 1974 technological improvements in the smelting industry have been brought about because of the concern about fluorine affecting pastures and animals in other parts of the world so that a closed potline system is now available.

This means that the 3,000 kilograms of fluorine ejected daily from a tall chimney need no longer be dispersed at all. Fluorine can be trapped at the potline and tall chimneys have been superceded by dry scrubber systems.

This new system has been advocated by at least one of the smelter companies known to be interested in Aramoana. Otago authorities must insist that this system is installed if Aramoana is to have a smelter.

This system must also involve satisfactory arrangements for disposal of the fluorine elsewhere than Aramoana. This matter would become an important point to discuss when firm proposals come forward.

BUILDINGS AND SERVICES

Buildings at Aramoana to house the smelter would require an overall area well within the 80 hectares available....

Buildings would be substantial but not tall since the potlines require single storey structures....

At this stage it appears that a buffer zone of native shrubs and trees could be planted between the dry sandy zone which offers the desirable industrial site and the tidal flats which provided the wildlife zone.

In 1974 discussions on the nature

of loading and unloading facilities took place and it was made clear that massive embankments built across the sand would interfere seriously with the biololgically rich zone.

It became clear that it would be possible to build the wharf required....

Several services would need to be brought into Aramoana. Water supplies are available through the City Council's existing system. ... A smelter can recycle its water so that there is no need to discharge hot water into the harbour. This should be insisted upon in any discussions with potential users.

Electricity pylons would be needed to bring power from the national grid. Power line routes are always worked out with the close advice of the Nature Conservation Council and its responsibility would include detailed consideration of the wildlife using the tidal flats at Aramoana, as well as the wellbeing of the albatross colony two kilometres away.

If necessary the last kilometre or more of these high-tension cables could be put underground as they approach the smelter site and any sub-station could be built further away.

Road links to the site are certain to be needed with rather better

facilities than exist at present but there should be no need for a massive highway since the principal materials needed at the site would come and go by sea.

The present road winding around the bays from Port Chalmers would need some reshaping....

We now know that such bays need satisfactory daily flushing by the tide if they are to remain attractive and biologically sound. This would mean putting more of the road across each bay on pillars at the head of some of the bays and avoiding long embankments that cut off the free flow of the tide.

Some mention has been made of the need for housing at Aramoana which is a rather windy and cool site at the best of times. I see no need for any settlement at Aramoana apart from a small number of supervisory or custodian staff. The workforce could live elsewhere and come to the site to work.

The Peninsula bays across the harbour provide far more satisfactory living conditions with their north facing aspect, ... while a number of people would no doubt prefer to live in Dunedin where town planning investigations have shown that there is under-utilised capacity already fully serviced for considerable numbers of additional homes....

5.   Can environmental problems

be resolved?

Given present-day technology, the picture is very different from the situation that would have resulted from any smelter at Aramoana some years ago.

The Wildlife Service has itself investigated the Aramoana tidal flats and appears ready to recommend that a wildlife reserve should be declared and that it is, in their view, possible and sustainable.

The Lands and Survey Department has pressed for the coastal strip at Aramoana to be reserved for recreational purposes....

The whole question of dangerous fumes can be avoided by insistence on a smelter with no chimney and the disposal of waste products must be worked out before the site can be accepted as satisfactory. There is no reason to suppose that this cannot be done.

'Buildings Would Be Substantial But Not Tall'

These considerations lead me to believe that it is possible to consider a major aluminium industry on the site at Aramoana using both its first-rate industrial potential and respecting the high quality wildlife values and the recreational potential of the coast....

SOURCE: *Otago Daily Times*, 28 April 1980.

The problem of where to locate a smelter is a universal environmental issue and students should consider its desirability and feasibility by referring to the sketch map, Figure 4.11. Hypotheses covering desirability and feasibility can be formulated before reading the article. From Professor Lister's article, points on desirability/undesirability, feasibility/unfeasibility can be extracted. A values analysis strategy parallel to the battered wives activity can then take place or a procedure set out in McConnell *et al* (1979) for studying local controversial issues may be preferred (see Figure 4.12).

The broad meaning of data

Question-asking activities in learning, together with reaching generalisations and decisions through data

processing have again been dominant themes in this chapter. It is clear by now that data are given a very broad meaning to include not only statistics and evidence gleaned from reports, maps, photographs or newspapers in the scientific tradition, but also the reactions and feelings, ideas and thoughts of students confronted with humanistic dilemmas and problems of values and attitude analysis. Such a broad definition of data is necessary if the student's own knowledge and feelings are to be valued, developed and transformed in an educationally worthwhile undertaking.

As a template of the teaching-learning process the question identification, data examination, development of understanding procedure, seems to have validity and mirror the reality of planning for teaching

Figure 4.11 Choosing where to locate a smelter

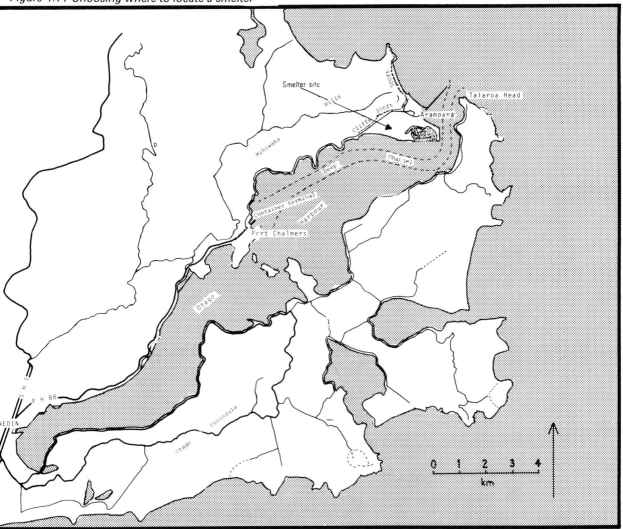

and learning. The examination of attitudes and values, no less than the development of understanding and skills, may be undertaken using the question identification, data examination, decision resolution plan.

Final suggestions

As a series of final suggestions in values and attitudes teaching in geography, I suggest the following as a checklist based on ideas from Watson (1977).

1. Select an issue with which students can identify. You want to engage their feelings.

2. Use a case study which is contentious and for which there is ample background information. Students must be able to identify from data the range of value/attitude positions manifested in the issue.

3. Classify the various value/attitude positions and help students to analyse the pros and cons and implications of the range of positions.

4. Evaluate the arguments by comparison and contrast. Encourage students to relate their personal values to the values identified in the particular case study.

5. Involve students in making a choice, perhaps by ranking the range of priorities operating and then having them decide what their position would be.

6. Give students the opportunity to provide thoughtful and convincing reasons for their choice or decision through written reports, debate, simulated interviews.

Figure 4.12 Procedure for studying a controversial issue

Becoming aware of and clarifying the issue

What is going on here?
What is the major issue, problem or question?

↓

Analysing and expressing one's feelings

Why did I choose this problem to work on?
Have I made any assumptions about it?
What are my attitudes? My bias?
How do other people react to the problem?

↓

Inquiring, carrying out research and reaching the best possible factual judgement

What do I need to know? How do I find out about it?
What knowledge or methods are useful to me?
What have I found out about the problem?
What are the implications of these facts?

↓

Clarifying values and reaching a value judgement

What values and beliefs do I hold that are relevant to the problem?
What are the consequences of these in relation to the problem?
When there is conflict between values, which values have priority?
Why?

↓

Synthesising fact and value judgement and making a decision

Do I have enough information to decide?
If not, what do I do?
What solution do I now propose?
What courses of action are open to people wishing to bring about the solution? Should I become personally involved? If so, how, to what extent and why?
What are some of the possible consequences of my involvement or non-involvement? For society? For myself?

↓

Doing and evaluating

What did I actually do?
What were the consequences of this?
Knowing what I know now, how would I act if the situation arose again? Why?

SOURCE: McConnell, W. F. *et al* (eds) (1979) *Studying the Local Environment*, Allen & Unwin.

Where do you stand?

And if you wish to assess your likely position in relation to taking up values teaching strategies in geography, answer each of the following questions with a yes or no:

1. Do you want to help students *examine* their personal feelings and actions in order to increase their awareness of their own values?
2. Do you want to stimulate your students to develop higher forms of *reasoning* about values?
3. Do you want to help your students use *logical thinking* and *scientific investigation* to analyse social value issues?
4. Are there certain *values* and *value positions* that you want your students to adopt?
5. Do you want to provide definite opportunities for your students *to act* individually and in groups according to their values?[3]

Further reading

Allen, R. A. (n.d.) *But the earth abideth forever*, National Council for Geographic Education, Instructional Activity Series 1A/5–16.

Blachford, K. R. (1972) 'Values and geographical education', *Geographical Education*, Vol. 1, No. 4, pp. 319–29.

— (1979) 'Morals and Values in Geographical Education Towards a Metaphysic of the environment', *Geographical Education*, Vol. 3, No. 3, pp. 423–57.

Cowie, P. M. (1978) 'Geography: a value laden subject in education', *Geographical Education*, Vol. 3, pp. 147–61.

Fien, J. and Slater, F. (1981), 'Exploring values and attitudes through group discussion and evaluation', *Classroom Geographer*, April, pp. 22–5.

Graves, N. J. (1979) 'Contrasts and Contradictions in Geographical Education', *Geography*, Vol. 64, November, pp. 259–67.

Hall, R. (1978) 'Teaching humanistic geography', *Australian Geographer*, Vol. 14, May, pp. 7–13.

Huckle, J. (1976) 'Values and attitudes—the geography teacher's new frontier', *Teachers Talking*, April, Thomas Nelson.

— (1980) 'Classroom approaches: towards a critical summary', in Rawling, E. (ed.), *Geography into the 1980s*, Geographical Association.

— (1981) 'Geography and values education', in Walford, R. (ed.), *Signposts for Geography Teaching*, Longman.

Kracht, J. B. and Boehm, R. G. (1975) 'Feelings about the community: using value clarification in and out of the classroom', *Journal of Geography*, Vol. 74, No. 4, April, pp. 198-206.

— and Martorella, P. H. (1970) 'Simulation and Inquiry Models applied to the study of environmental problems', *Journal of Geography*, Vol. 69, No. 4, May, pp. 273-8.

Martorella, P. H. (1977) 'Teaching geography through value strategies', in Manson, G. A. and Ridd, M. K. (eds.), *New Perspectives on Geographic Education*, Kendall/Hunt, pp. 130-61.

Smith, D. L. (1978) 'Values and the teaching of geography', *Geographical Education*, Vol. 3, pp. 147-61.

Wolforth, J. R. (1976) 'The new geography—and after?', *Geography*, Vol. 61, July, pp. 143-9.

5 Learning through Geography

A procedure for planning learning activities has been outlined in the preceding chapters. It must necessarily be only one of a number of approaches to lesson planning and geography teaching as teachers' preferences and views of themselves, a school system's objectives and demands, student motivation, and societal pressures obviously differ widely. As a procedure, it identifies a sufficient number of steps to enable it to be used as a sound framework for guiding planning and learning. A desire for structure and planning in some sense are, of course, necessary attitudes to bring to the task.

Basic assumptions

I shall now touch briefly upon some of the assumptions which I think I have made by advocating a question identification approach to planning and then widen discussion to point up the strengths and possible weaknesses of the procedure.

Certainly, I am assuming that students have made the social adjustments necessary to working and learning in a class with twenty or thirty other individuals. Such social adjustments need not be listed at length but include a willingness and predisposition to learn in that setting, to respect the rights of other individuals and the leadership role of the teacher. It is taken for granted that in terms of Maslow's (1943) hierarchy of needs, that the so-called 'lower order' needs have been satisfied.

I am also counting on certain levels of motivation and a capacity for self-direction amongst the students, of which one would be the ability to come to lessons prepared with paper, pencils, books and so on. Such a lack might not prevent an activity approach to planning but it would make the undertaking more difficult and less easy to get underway and sustain.

Michael Storm (1979) has recently, quite rightly, pointed out that discussions about teaching (and learning, I would add) are dominated by considerations of content and methodology with management

aspects being apparently taboo. He argues that it is futile to say that if content is right and teaching methods stimulating, control problems disappear and social adjustment is automatically attained. Lack of control and positive attitudes to school make it difficult and perhaps impossible to introduce a range of materials or methods into a classroom. An extensive data base on which to direct questions to work towards generalisations requires such a range of materials and methods. Figure 5.1 expresses the idea that a lively new activity results in disruption. A teacher reverts to routine tasks known for their management effectiveness (though sometimes the apparent futility of such tasks can produce its own behaviour problems). Where management is a dominating concern, teaching for generalisations and decision making is not likely to be a viable priority since it requires active class participation and sustained teacher-student interaction and co-operation.

It is clear also, that I am assuming that a teacher is

Figure 5.1 Hazards to introducing new teaching methods

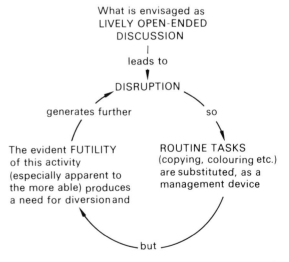

SOURCE: based on Storm, M. (1979) 'Some tentative thoughts on a taboo topic', *Geography Bulletin*, I.L.E.A., No. 6.

not only able but willing to adopt this approach to lesson planning and this may not always be so. Willingness to consider changes, even quite small changes, in teaching style requires a degree of open mindedness and propensity to experiment that raises anxiety levels, calls for fresh preparation and puts perhaps a smoothly working system at risk. However, the need to change, to innovate, to introduce new teaching strategies as well as content and to take a fresh look at one's planning procedures is probably a professional obligation. (Hickman *et al*, 1973.) This may be taking for granted, however, that all teachers see teaching as the same kind of undertaking.

The teacher's perception of the task

It is likely that to the teacher surviving in the hostile classroom very different preoccupations and values will be held from those working in other conditions. Reflection on styles of planning activities could well be a superfluous luxury as stamina and morale building take precedence over contemplation and organising learning as opposed to containing learners. But, not all teaching is teaching for survival and indeed such a concept of teaching may only be applicable to the early experience of teaching though such an arduous initiation may close minds to activity based learning for a long time.

In a normative sense, teaching is a profession, even, I would add, a caring profession. A teacher has an obligation to attempt to move a class along from hostility, towards acceptance of school and education in an active sense. Such an obligation follows from one of the defining attributes of professionals—their commitment to an ideal of service, in this case service to education. And necessary and valuable as social control and shaping might be, it cannot stand in permanently for education in a complete sense.

Professionals committed to education are obliged then, to act, themselves, as educated people and to reflect on their practice and the implications of this for their students who are in the process of being educated. I have been assuming within my structured planning model that quite a rich learning environment can be provided by teachers and accepted by students. A school system's and society's demands, as well as teachers' and students' personal attitudes, may preclude and/or reject this. There may be in all four cases good reasons.

The planning procedure

To return now to a recapitulation of some of the features of the planning procedure. The structure of the procedure quite clearly lies in the specification of a number of steps to be followed of which the most crucial are:

1. Listing a set of questions to form the enquiry sequence.
2. Deciding on learning activities and teacher strategies.
3. Gathering and developing resource materials to help answer the questions.
4. Selecting methods of data gathering and processing.
5. Organising learning and teaching to reach generalisations.

Following these steps achieves an *organisation* of classroom activities. It does exclude leaving students to their own devices in what might be two extreme situations. For instance, students could be given a text and told to learn from it or, alternatively, they could be given a range of materials and told to sort out answers for themselves. Some people, usually very mature people, learn a great deal from such unstructured, undirected, extreme descriptions of teaching, though I doubt that such caricatures contain much which conceptualises teaching as an activity carried out for the benefit of growing adults. The planning procedure outlined in this book gives students a direct entry into investigation through questions and into a form of analysis through resources provided for processing.

To what extent does the plan ensure that teacher direction and student involvement are in balance? To what extent could the following lesson plan be the result of planning using the question identification approach?

1. Read passage from textbook.
2. Teacher *tells* students the main points.
3. Teacher perhaps expands on these and corrects any confusions or omissions.
4. Students write notes as a summary.

No enquiry sequence has been planned in which the students participate in more than perhaps a minimal fashion. No generalisation is being explicitly *linked* to questions. The data are already in the form of answers.

Truncating teaching

This teaching strategy lacks the spirit and purpose of the questioning approach and also most of the procedural steps, though it could be argued others are there—data and a method of data processing are identifiable. Clearly, such a strategy is, in my terms, a severely truncated learning experience, and it is not akin to the question identification procedure.

Students may not have any idea of what overall question, or problem, 'the main points' relate to or indeed how the main points relate to one another in a general way. There is no enquiry sequence to guide and inform.

I would argue that teaching methods similar to the above do not provide sufficient guidance for learning and developing understanding—except at the very lowest of levels—they are over-condensed and closed procedures on which to plan for learning. The textbook is the data source and the chief objection here is that the information is most likely to be presented in finished form though this need not be so and increasingly is not the case. The students are not using the data to reach their own conclusions but rather to summarise the conclusions of others. Good students may go beyond this to make connections for themselves but not all are being helped to do this. Then, too, the interactions between teacher and students are, on the basis of the descriptive points given above, likely to be limited to telling what has been comprehended and analytical skills are limited to notemaking. This is not completely and utterly to decry comprehension and note making as data gathering and processing methods but to suggest that such activities are only two of a tremendously wide range of skills which can be developed and used in the process of learning.

Guiding learning

One of the essential differences in emphasis between the questioning approach and the scenario lies in using data and data processing methods not to make students rediscover all that is already known but to guide them towards seeing patterns in data and formulating relationships. In the exercise on settlement size and communications in the Shepparton district for example, the relationships articulated depended on seeing patterns in the data.

To give opportunities to examine data as evidence and to draw conclusions seems to be a key distinction between what some would call expository (telling) teaching and enquiry teaching. That enquiry style teaching requires some exposition from time to time goes without saying, though sometimes appears to be forgotten.

If the planning procedures outlined in this book ensure in any way that the reading–note-making stereotype of classroom learning is expanded, then it will be through an appreciation of the crucial importance of identifying questions and generalisations in planning for learning to develop understandings.

Enquiry sequences could be set up to guide students through a written account in a text towards a generalisation. With questions specified and a generalisation to be developed, the set reading becomes more meaningful even if students are working on finished product data.

More guidance is provided for learning than in the above scenario and it is the guidance provided for learning which I see to be one of the strengths of the procedure. It allows for the connection between question and answer to be maintained and to be maintained in the process of moving from one to the other during an examination of data. It also needs to be said that the plan is sufficiently general to leave many decisions to the teacher and herein lies a degree of flexibility. The questions and concepts, the data and data processing methods, the generalisations, even the particular view of geography, are left open to the classroom teacher.

A guide to learning

The claim that a question identification approach to planning is also a guide to learning rests on the logic of the idea that knowledge consists not only of answers to questions but of the questions and answers taken together. Others claim to identify other ways of meshing 'doing' geography with learning geography. One of the hopes to emerge and subside during the conceptual revolution was that a parallel structuring of concepts in the discipline and structuring learning would be found.

Clegg (1969), using Taba's (1962) cognitive process model, has made a reasoned case, based on a field work activity, for combining the process of learning with the content of geography into a number of pedagogical principles. In a field study of the land use of Amherst, Massachusetts, the Central Business District (CBD) and some outlying neighbourhoods were mapped. The generalisations being worked towards, were stated as follows:

1. As a result of making actual field observations in a selected town and organising the data on a series of transparent overlay maps, students will be able to prepare a brief report:
   (a) identifying land uses;
   (b) analysing land uses to determine what patterns of distribution exist;
   (c) developing a theory about the function of a town based on its patterns of land use;
   (d) predicting changes likely to occur by 1975 given the projected growth rate of the town and its major components during 1968–75.[1]

Content and concept

In order to show how content and concept development was combined, the product and process steps are outlined in parallel columns, reproduced here in Figure 5.2. The collection of information on land use parallels the development and identification of concepts. As the data requires classification, so students label and group items—a classic step in concept development. In the summary and analysis of data within a data retrieval chart (Figure 5.3) concepts of clothing and hardware stores are categorised into the more generalised concepts of consumer products or services, for example. Then references and conclusions are drawn from the table, e.g. Clegg reports such generalisations as:

1. Auto services are clustered in the central business district;
2. Gas stations provide special services to small residential areas for the convenience of people living there;
3. Auto services are dispersed on the fringe areas, which are connected with the CBD by the main entry and exit arterials.[2]

Comparisons made by reading across the data retrieval chart were also encouraged and relationships realised.

Finally, in order to *apply* the principles or generalisations stated, attempts to predict future patterns were attempted. As data was gathered and processed, concepts and new relationships or generalisations formed. The development of the study paralleled the development of thinking. This is probably easiest to do in a field study though given appropriate secondary data the procedure could be equally well applied.

Both my view and Clegg's views of planning for learning represent attempts to assist learning even if they are very teacher-centred perspectives and views of how learning takes place i.e. teacher-centred in that we are both operating on hunches as much as on any overall theory or absolute proof. I have argued for connecting questions and answers. But apart from our knowledge of the psychology of learning, and this is not always easy to translate into conscious classroom practice, we actually do not know a great deal about the formation of thoughts and ideas and *how this is best assisted.*

Language and learning

The relationships between language and learning has been an acknowledged, if largely unexplored, field until recent years. While the theories and ideas, definitions and concepts of Piaget, Bruner, Gagné and Ausubel have influenced practice (to mention only four psychologists well-known in geographical education), it seems that the message of research workers on the role of language in learning is less widespread and certainly less sympathetically received. This probably arises partly from confusion of the two functions of language.

Language functions chiefly in classrooms and elsewhere to communicate what has been learned and is known, and yet it is also part of the *activity* of learning. Both functions need to be recognised and the latter given some more attention (DES, 1979). The former undoubtedly receives most attention at present and while few would decry an ultimate need for the correct and expert use of language in communication, we need to give more opportunities for the penultimate need provided by the process of 'talking to learn'. 'Talking to learn' assists in understanding technical terms and ideas, for example.

A more learner-centred view of learning and the role of language has been put forward by the Schools Council's *Writing Across the Curriculum Project* and others. Their message, though not without its critics (Williams, 1977), needs to be considered when planning learning activities. It is not sufficient to follow through procedures for learning, to select and devise learning activities for students—learners need, as

Figure 5.2 Linking cognitive processes and learning geography

<table>
<tr><td>

THE PROCESS AND PRODUCT OF LEARNING GEOGRAPHY

I. *Micro-study of a Town*

A. Make field observations of different kinds of stores, public buildings, unused land, etc. Record each item separately.

B. Sift through the recorded items. Make tentative groupings according to common characteristics. Shift some items as new groupings become evident.

C. Decide upon an appropriate name or label for the groups that best reflects the basis on which the grouping was made. Arrange groups and subgroups into hierarchies of functions within the broad groups or categories developed.

II. *Summary and Analysis of Data*

A. Develop a data retrieval chart (Figure 5.3) for summarising data and for recording answers to analytical questions to be asked about the data.
1. How are the land uses *grouped* or *clustered*?
2. How are they *dispersed*?
3. What is the *shape* or *pattern* of distribution of land uses?

B. Develop summarising statements for each grouping by reading down the columns of the data retrieval chart. Make comparisons and contrasts between groupings by reading across the rows of the chart. Show relationships among items. Develop explanations. Point out limitations. Draw conclusions.

C. Recognising that the data are confined to only one sample, make tentative inferences about the meaning of the data. Develop statements or generalisations that go beyond the actual data and suggest conclusions of wide applicability. Develop charts, maps, or graphs to summarise or illustrate these conclusions.

III. *Application to Current or Future Problems*

A. Obtain population projections for 1968–1975 from the town planning office or university planning offices. Estimate the number of new services that will be necessary to meet increased populations. Predict new locations or distribution patterns for such services. Predict new services that are likely to develop. Predict whether the present patterns are likely to remain constant by 1975.

B. Adduce logical arguments to support the predictions. Cite evidence of trends in Amherst, or in other towns that would lend support to the prediction.

</td><td>

COGNITIVE PROCESS AND TEACHING STRATEGY

I. *Concept Formation*

A. Observation and gathering data. Enumeration and listing.
Teacher asks: 'What did you see? Note?'

B. Grouping by identifying common properties; abstracting.
Teacher asks: 'What belongs together? On what basis?'

C. Naming, labelling, and categorising. Determining hierarchical order of items.
Teacher asks: 'What would you call these groups? What belongs under what?'
Note: The name of the category or the concept is an abstract term that is useful for categorising phenomena or events. The concept usually takes the form of a single word or short phrase.

II. *Interpreting Data, Making Inferences, and Developing Generalisations*

A. Identifying and differentiating data according to the conceptual categories developed above.
Teacher asks questions such as those in the left-hand column.

B. Explaining items; relating points to one another; seeking possible cause and effect relationships; and recognising limits of data.
Teacher asks: 'How can we summarise all this information? How can we explain it? Why did these groupings occur? What other explanations are possible?'

C. Making inferences: going beyond what is given. Finding implications, extrapolating. Drawing conclusions, forming generalisations.
Teacher asks: 'What does this mean? What would you conclude? How would you say it so that it would apply to many similar situations, not just this one?'

III. *Application of Principles*

A. Predicting probable consequences of events or courses of action.
Teacher asks: 'What would happen if . . . ?'

B. Explaining, supporting, and verifying the predictions.
Teacher asks: 'Why do you think this would happen? What would it take for the prediction to be true?'

SOURCE: Clegg, A. A. (1969) 'Geography-ing or doing Geography', *Journal of Geography*, Vol. 68, No. 5, National Council for Geographic Education.

</td></tr>
</table>

Figure 5.3 Categories of land usage

	LAND USAGE						
Analytical Questions	Personal services	Consumer products or services	Food services	Profes- sional services	Public services	Auto services
Where located? How clustered? How dispersed? Pattern discernible?						

SOURCE: Clegg, A. A. (1969) 'Geography-ing or doing Geography', *Journal of Geography*, Vol. 68, No. 5, National Council for Geographic Education.

well, opportunities to make meaning through talking and writing, to use language as a way of cultivating learning. The role of language needs further description and explanation.

Talking to learn

In response to the question, 'What is a river?', A. J. Lunnon (1979), in a thesis designed to ascertain the understanding of selected concepts often used in geography at the primary school level, lists the following responses from pupils in the 8–13 age range:

It is water
It is blue
It is a sort of stream
It is like a lake

A second question, 'Why do you say "like a lake"?', designed to probe the last response, elicited the answer, 'Well, it's not got so much water in it'. Other responses included:

It is water which moves
It is a kind of big stream which leads into the sea
Like a pond but running
It is a lot of rain water which leads into the sea
Flowing water in a long thin ditch
A big stream which wanders in and out of the hills

These replies constitute an example of what the *Language Across the Curriculum Project* team, working at the University of London Institute of Education in the first half of the last decade, would classify as 'learning in transition'. An integrated concept of the many variables which characterise rivers has not been attained in any of the above definitions *but*, to a greater or lesser degree, each represents an *attempt* to organise ideas about rivers.

Presumably, as a result of being questioned, each child realised something of what he or she could and could not put into words about rivers.

From Lunnon's work, we have no further evidence of the concept 'river'. (This was not the purpose of his research.) Yet, it would probably be safe to assume that every child had used the word *river* many times in verbal and written communication. Lunnon's evidence of concept attainment suggests that children need many opportunities to explore the meaning of words and to grapple with the task of catching hold of several ideas and moulding them into a concept.

From information to understanding

The Project, in the course of its work, was able to produce many examples of the learning which can take place when children are presented with the opportunity to talk through concepts and ideas. The examples illustrate very convincingly how, through talk, children clarify their ideas, come to realise what they do not understand and yet work through what they do know to make new connections.

The examples are too long to reproduce here but one particularly striking one is to be found in the

Project bulletin *From Information to Understanding* (1975, 3–16) (see also Slater and Spicer (1980)).

Talking through concepts and ideas is classified by the Project as expressive talking or writing, and they believe that such exploratory opportunities are vital to learning. The development of concepts would seem to be not solely dependent on maturation but also on the active use of language—talking and writing. The learning process is about clarifying and expanding concepts.

To be more aware of, and sensitive to, the role of language in learning is the challenge which the Language Project threw out to subject teachers. Should we consider when planning learning activities what opportunities are being made for expressive talking and writing? Can we be persuaded to see such episodes as learning activities, as another form of data processing, perhaps the most fundamental form of data processing?

In Piaget's terms, we need to consider that expressive writing assists in the effective assimilation and accommodation of new ideas into existing mental schemes; that new learning has to be linked up to previous learning and understanding; and developed from there to make meaning. The Project found that in geography lessons in England and Wales, talking and writing appears to be in the transactional mode 99 per cent of the time. Would this be true in other educational systems? What do *freer* writing opportunities reveal?

Examining language

It may be possible to select several pieces of written work produced in one of your geography classes and to examine the work in relation to the following questions:

1. To what extent is language being used to explore ideas or relationships? What do you think has caused the child to write in this way?
2. To what extent does the language demonstrate that the writing has been an exercise for learning? How much is likely to be the student's own thinking?
3. What evidence is there that a reshaping of ideas is taking place? What do you think is the main purpose of the written work?
4. What has the piece of writing done to move learning forward? To whom does the writing seem to be addressed?

5. Do you feel that what is written constitutes acceptable language? Is there evidence to suggest that learning has been moved further forward?

Stages in the use of language

At this point, it is necessary to offer definitions of expressive, transactional and poetic writing to assist in an analysis of writing or talking in a geography lesson. The definitions are not meant to be taken as definitive. The three Project definitions are related to one another thus:

Transactional $<---$ Expressive $--->$ Poetic

Transactional writing, or the language for getting things done, is writing or talking to inform people, to advise, persuade, or instruct people. Transactional writing is used, for example, to record facts, exchange opinions, explain ideas, construct theories, transact business, conduct campaigns, change public opinion. Transactional writing is passing on accurate information in an ordered sequence.

Expressive writing is the kind of writing that may be called 'thinking aloud on paper'. Expressive language is language close to the self and it has the function of revealing the speaker, verbalising his or her consciousness and understanding. In expressive speech or writing the person feels free to jump from fact to speculation, from personal anecdote to emotional outburst, none of which will be taken down and used in evidence against him or her. It is the way in which new ideas are tentatively explored, thoughts half-uttered and half-expressed. The rest is to be picked up by a listener or reader who is willing to take the unexpressed on trust. Expressive writing is a presentation of an experience as recalled—'warts and all'—or of ideas in the process of clarification.

Poetic writing is a shaping of words into a form for their own sake. The function of a piece of poetic writing is to produce an object that pleases or satisfies the writer and the reader's response is to share that satisfaction.

Talking and writing in geography lessons

It is very likely that transactional writing and talking dominate most geography lessons and that teacher talk to student talk is in the ratio of at least 3:1. Furthermore, the teacher will be using well-formulated transactional language, the concepts and ideas being presented in a highly articulate polished fashion. Little opportunity may be given for students to explore meaning for themselves. This is scarcely surprising since little emphasis is given to the role of language in understanding beyond primary school.

The point which the Language Project came to consider as their central argument is that a chance to use expressive writing and talking is essential if understanding is to be reached. The demand for impersonal, unexpressive writing can actively inhibit learning because it isolates that which is to be learned from the vital learning process—that of making links between what is already known and the new information. It is through the tentative, inarticulate, hesitant, backward- and forward-moving, expressive mode that connections and links between old and new knowledge come to be made. Then a student may be ready to set the understanding down in a formal transactional mode. Expressive writing is considered to be the seed bed from which more specialised kinds of writing grow. The writing may move towards the greater explicitness and clarity of the transactional or the conscious shaping of the poetic. One other essential point: in expressive speech or writing we put ourselves in a trusting relationship, that is, the teacher or fellow student is not seen as marker or critic but as guide, helper, or just as sympathetic listener.

Encouraging the expressive mode

The team suggested a two-fold policy for encouraging the expressive use of language, the use of language where people speak aloud or jot down thoughts in attempts to make connections and forge new understandings of concepts and ideas. They advocated (1) providing a variety of audiences for the spoken and written work of students so that there is a decrease in the number of times the teacher is seen as an examiner evaluating whatever is written, and (2) giving students a range of writing purposes so that they are allowed the chance to express their thoughts in a variety of ways.

Audiences which would provide a range of purposes and make more room for expressive language as an essential stage on the road to understanding include:

1. Student to self, as in a diary of my journey to/ the visit to ...
2. Student to trusted adult, for example, 'What I learned from the film/the field work at; What I see on this map/in this photograph ...'
3. Student to student as partners in a dialogue, for example, discussing a particular concept, participating in a game. Games seem to offer wide opportunities for using language informally and at the expressive level. The debriefing discussion with the teacher can also be at the expressive level and apart from setting up buzz groups to discuss concepts such as resource or habitat, debriefing after a decision making activity is the only time when I can recall giving my geography classes in New Zealand the chance to talk informally for any length of time. Once I began to learn of the Language Project I realised just how few chances they had had to explore ideas for themselves. (I think that the pattern I followed in English lessons was very different where discussing a play, a poem and so on demanded 'warts and all' dialogues.)
4. Student to teacher in rough drafts of field reports or essays where comments are made to assist, diagnose strengths and weaknesses, suggest improvements before a final draft is marked for examination or test purposes. From rough to final draft represents a move from expressive to transactional which is also true of 5 below.
5. Student to student writing when position papers or speeches are being prepared. Alternatives 3 and 4 also provide opportunities for collaborative writing.
6. Writer to readers (relatively unknown audiences), including the design of advertisements, preparation of letters to newspapers or town planners. In such tasks language is likely to move from the expressive to the transactional.

As a final point, it should be emphasised that the message of the curriculum project outlined here should not be dismissed as 'nothing more than saying that every teacher is a teacher of English'. Rather, every teacher is a teacher of understanding and language in its exploratory, hesitant, experimental, pause-peppered 'er-um' phase and is an essential building block as 'learning in transition' moves to learning as understanding. In my opinion, the more humanistic kinds of exercises such as, 'What do you like/dislike about a place?' give opportunities for 'freer' talking and writing. It takes more thought and nerve to introduce planned opportunities into geography as science undertakings.

Monitoring what is being learned

One of the anxieties and uncertainties that arises for the teacher when putting students into discussion groups or giving them opportunities for expressive talking and writing simply hinges on whether anything is being learned. We rarely have the time to take transcripts and make tapes and listen to them in order to obtain some reassurance or indeed some knowledge of where students start from and move to. Another skill which it takes time to develop is the ability to interpret other people's talk from the viewpoint of the meaning being constructed and knitted together. Douglas Barnes (1969, 1976), provides a number of transcripts of direct interest to a geography teacher and through these and his interpretations one can begin to get some feel for the role of language in learning.

Saxons and settlements

The transcript reproduced in Figure 5.4 with Douglas Barnes's commentary, is of a group discussing, 'What would a Saxon family first do when they approached English shores in order to settle?' The discussion took place as part of a history lesson and Barnes considers that the students were already well provided with background information. They had a paragraph and sketch in front of them. The task before them is to organise and relate old knowledge to the present task. The group seems to be doing this quite successfully as a number of significant items concerning settlement are identified. Barnes considers tentativeness to be particularly valuable in keeping discussion open and flowing. Questions are interpreted as tentative suggestions and words like 'say', 'probably', 'I suppose', function as encouragement to develop ideas further.

Barnes argues that the rearticulation of knowledge is enabling things like wooden houses, clearing timber and shelter to be interrelated and given new meanings in relation to the question of siting a Saxon village. The settlement siting activity outlined in

Figure 5.4 Saxon settlements

	Dialogue	Commentary
28.B	The Saxons used er timber didn't they to ...	Betty begins the sequence with what at first glance appears to be a statement. It functions however as a hypothesis inviting further exploration. (Implicitly: How should we take this into consideration in choosing a site for the village?)
29.	Yes	
28.B	[cont.] ... to build houses?	
30.T	They cleared a ... Say they found a forest and you know they're probably all forests near the ... [inaudible]	Theresa takes up the implicit suggestion of the need for a site with a plentiful supply of timber. The 'Say' formula and the 'probably' invite the others to regard this contribution not as final but as open to qualification.
31.B	Yes. They cleared it all away ... and then built all the little huts and brought all their animals and ...	Betty accepts the invitation and develops the idea further.
32.C	... All the family and that. They'd have to be pretty big huts.	Carol has not been following this line of thought, and now interrupts Betty with a dogmatic assertion which could lead in another direction.
33.T	Yes.	This is politely acknowledged but taken no further.
34.B	Why did they live in valleys?	Betty rescues the group from the dead end by raising a new question (provoked by the textbook illustration).
35.	[Long pause] Aarh.	
36.T	I suppose so ... so they ... they'd be sheltered.	The tentativeness with which Theresa eventually offers an answer is expressed both by her hesitations and by 'I suppose ...'
37.B	Yes, for shelter ... and so er ... so there was less risk ... of being attacked I should think.	Betty accepts Theresa's answer but puts an alternative one of her own beside it; her hesitant delivery and the phrase 'I should think' disclaim any pretension to firm knowledge and implicitly invite further additions or qualifications.
38.T	Yes.	Message received.
39.C	Because they could only come from two directions.	Carol accepts the invitation and extends Betty's suggestion a step further.

SOURCE: Barnes, D. (1976) *From Communication to Curriculum*, Penguin Education.

Chapter 1 could well be organised as a group activity like this. The rearticulation and reinterpretation of previous learning replaces what might otherwise be a period of teacher elaborated recapitulation. Barnes and others would see it as essential to give students the opportunity from time to time to talk themselves through such recapitulation of knowledge which in some senses they already possess.

Another section from the same group, but accompanied by a diagram to show the line of thought (see Figure 5.5) helps to convince that discussion can be connected and purposeful. It needs to be appreciated

Figure 5.5 The structure of thought in a discussion

4.C	When the boat lands the first thing they'd have to do ... be ... to find ... em place where they can build a house, and probably later on have ... fields of their ... crops and ... places ...
4.C[cont.]	... to keep ... em ...
5.B	They'd probably look round first.
4.C[cont.]	... cattle and [inaudible] ... pigs and things.
6.T	But they'd have to be out of the way of swamps and things ... so they wouldn't be in any danger.
7.B	You could say that when they arrived there they wouldn't use the ... em ... Roman things ... that had already been put there.
8.T	They wouldn't go near them because they were scared of the old Roman villages.
9.C	And ... th' ... they would p-probably ... keep away from the ... Roman towns erm ... the ... temples and that w–were mysterious and frightening places to them ...
10.C	And ... they were em ...
11.T	Yes.
10.C[cont.]	... they didn't understand things like that.
12.B	Yes ... probably they were ... an ...
13.C	And they've got to take ...
12.B[cont.]	... under-educated. [Amused]
13.C[cont.]	... and they've got to take care of all their animals and things because ... if they ... went too far from home they could die of exposure.
14.B	All that was in their old land, wasn't it?
15.C	No.
16.	Hmm ... [Long pause]

SOURCE: Barnes, D. (1976) *From Communication to Curriculum*, Penguin Education.

that, although conversation may appear to be jumping from one thing to another, it is not necessarily, in Barnes's word, 'shapeless'. The discussion focuses on what the Saxons would need—a place to build which was dry, safe and at a distance from previous Roman settlement—and what they would have to do—take care of their animals.

More complex ways of classifying conversations have been developed and it would seem that we should learn to see the logical process and the social skill components (Barnes and Todd, 1977) in group discussions with perhaps a view to identifying individual strengths and weaknesses. We could then begin to develop the *discussion skills* of students as well as allowing them some space in which to talk to learn. Talking or writing to learn should be seen, I think, as a data processing activity which leads towards general understandings which will be expressed eventually in transactional forms.

Relating ideas

The varying data processing abilities of students are apparent in transcripts—some students set up hypotheses, others make assertions, some put forward reasons and ask questions. Collectively, the group seems to be moving towards relating several ideas in order to make judgements about what Saxons would do in the initial stage of settlement. While all are

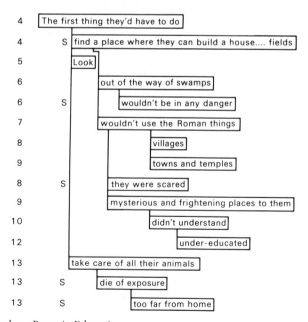

articulating ideas, we should note that there is no certainty, however, that each is combining a number of ideas. As well as relating ideas, the students are conceiving of possibilities beyond the limits of their immediate environment. These abilities—to relate ideas, to draw upon external reasons and ideas—are some of the features of the capacity to make a mature judgement in Peel's (1971) sense. It may help to gain some feeling of assurance that discussion is a productive activity, to recall that some students will be further on in the process of verbal reasoning than others. This fact merely *strengthens* the argument for recognising the role of language in learning and planning for language activity.

Language and logical thought

Rhys's work (1966, 1972) is of direct help in that it allows one to make evaluations about the level of thinking and judgement and quality of thought. A brief description of one of Rhys's studies will suffice as a reminder of (1) the description ↔ explanation characteristics or emphases in thinking often revealed, (2) the content-dominated nature of some thought through to (3) the possibility-invoking nature of other explanatory thought. A descriptive level of thought is characterised by an account of the event or phenomenon without reference to other ideas or possibilities. Explanatory thinking moves beyond content and immediate circumstances to introduce ideas outside the problem and data to hand.

Making connections

An aerial photograph of a small Canadian prairie town built around the intersection of road and rail was shown to secondary students. The students were asked individually, 'Why has this small town grown up just here, where the new road and railway cross each other?' The answers were classified into (1) content dominated, and (2) possibility invoking answers. An answer like this:

> 'So that the farmer can get his wheat harvest to the nearest town. The railway can carry much more than one truck and it's faster.'

clearly relies on the pictorial evidence and focuses on one aspect only. Another answer is very different:

> 'Because it is used as a central place with people bringing wheat to it to be stored and shipped off. It could be used as a central starting point and it provides a shopping centre. Also anything to be brought in for the farmer can be gathered here, and collected by the farmer later.'

This answer shows the ability to imagine possible factors outside the immediate evidence and these are co-ordinated to give an explanation of the function of the town. These characteristics of thinking should not be overlooked when the usefulness of talking through work is being questioned. Not all students will be displaying mature verbal reasoning abilities as they use language expressively—or in a transactional mode. This does not mean that having to talk about settlement requirements or functions is not helping the learning process.

I have emphasised the role of language in learning and the evidence and my own personal experience of how I learn is sufficient to convince me that using language in an exploratory way is part of the activity of learning. Others (Brent, 1978) attack Barnes for having a restricted and false view of meaning. However, from my reading of Brent, I consider that he fails to acknowledge that the emphasis and role Barnes and others give to expressive writing and talking is as a strategy along the way towards being able to express oneself confidently and clearly in the transactional mode. It may not be the *only* way to teach children the 'language game' of using publicly accepted meanings and methods of discourse but it may be the most humane way.

Using language to learn through geography is, as I have made plain, something one should constantly be aware of when planning activities and selecting teaching episodes. Yet to learn through geography requires the development of competence over a range of knowledge, skills and abilities. These have been specified in some detail in Chapters 3 and 4. To specify such knowledge, skills and abilities is to go some way towards a definition of what learning geography and learning through geography means. The planning model gives an overview of learning through geography. It raises an expectation that students participating in learning activities arranged in that style will be using questions as a guide to processing data to reach generalisations and decisions.

Planning for learning

This spells out not only a view of planning but, implicitly, a view of learning. This view has similarities with that set out for the social sciences by Piper (1976) as shown in Figure 5.6. His model is labelled as an enquiry view of learning.

Enquiry models, of which there are many, emphasise the importance of inducting students into procedures for carrying out at least some of the steps in an investigation rather than a view of learning in which the teacher tells, elaborates and explains and students comprehend and memorise.

Piper's model

Piper's model defines learning as a process of enquiry, development and growth. The process is held to involve the interaction of four areas of learning—knowledge, enquiry or intellectual skills, attitudes/values and social skills. It is not a psychological,

cognitive view of learning but an operational, functional view. There are a number of important distinctions between Piper's model and Bloomian models which should be noted in passing.

Knowledge is not confined to recall but includes understanding and realisation as well as remembering. The knowledge can be treated objectively as an ordered scientific kind or it can be experiential knowledge grounded in feelings and responses. Skills are defined as operational processes in the Piper model rather than mental processes. The emphasis is on their significance for handling data not on acquiring them as knowledge for its own sake. Significantly too, attitudes and values are not separated from the cognitive domain but are seen as part of it. This matches the view taken in Chapter 4.

A process model—the knowledge component

Each area of Piper's model is to be seen as (1) an input to a process and (2) an outcome of a process as well as being (3) an integral part of the enquiry-

Figure 5.6 Piper's enquiry model

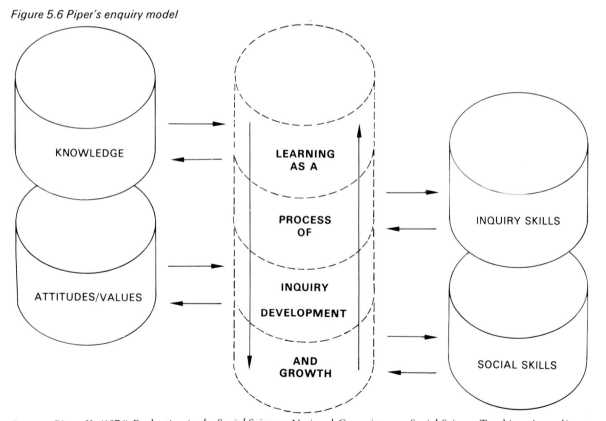

SOURCE: Piper, K. (1976) *Evaluation in the Social Sciences*, National Committee on Social Science Teaching, Australian Government Publishing Service.

learning process itself. Knowledge and skills have both a content and a process dimension. Knowledge as content is defined as information, concepts and generalisations, while knowledge as process specifies five levels of control over the data—recognition, recall, comprehension, application and realisation. Realisation is the ability to integrate new knowledge with old and to use knowledge creatively to generate original thought. So knowledge as realisation is viewed as something more than that first phase of generalisation and decision-making being worked towards in my procedural model for organising learning activities. It encompasses the use of that knowledge in subsequent thinking which I discussed in Chapter 2. To learn through geography requires using knowledge for remembering, understanding and realisation in a process of enquiry. These are the levels of operation in which knowledge should be engaged over a range of learning activities in Piper's very balanced view of enquiry learning.

The skills component

The area of enquiry skills is likewise defined by two dimensions, a context and a process dimension. The context derives from the form of the data: verbal, pictorial, quantitative or symbolic—convenient, if overlapping terms. The data, as we have seen in Chapter 3, may be presented verbally in books, newspapers, or interviews, pictorially in photographs or other illustrations; quantitatively as in statistical tables and graphs; or symbolically in maps and diagrams.

The stages in the process of enquiry are represented in Figure 5.7 which may be viewed as a grid for evaluating the range, variety and level of tasks which a learning activity requires of students. Within Piper's categories marginally modified here of (1) defining the investigation by identifying questions/issues/problems, (2) processing the data and (3) reaching and applying generalisations, eight sub-categories are developed.

Breaking down the categories

Sub-categories define even more closely what is involved in learning through enquiry and elaborate in detail the learning implied in my procedural planning model. The specific criteria in each stage are spelt out in Figure 5.8 and represent an amalgam of a selection of Piper's criteria with some very minor modifications I have made. Other useful specifications exist, for example the Liverpool Project's list.

The criteria are undoubtedly open to refinement and modification but serve here as a detailed summary of what might be expected to be some of the outcomes of a process of enquiry in the context of geography. Often the teacher will have organised the work through several stages and on other occasions, students will be involved at all stages.

Figure 5.7 The process of enquiry—the skills dimension

| Data Forms | Defining the questions, issues, problems | | Skills Processing the data | | | Reaching and applying generalisations | | |
	Identifying and clarifying questions and issues	Gathering and organising data	Interpreting data	Analysing data	Evaluating evidence	Generalising	Problem solving	Making value judgements
Verbal								
Pictorial								
Quantitative (i) Graphs								
(ii) Tables								
Symbolic (i) Maps								
(ii) Diagrams								

SOURCE: based on Piper, K. (1976) *Evaluation in the Social Sciences*, National Committee on Social Science Teaching, Australian Government Publishing Service.

Figure 5.8 The process dimension in enquiry

RESEARCHING THE QUESTION, ISSUE, PROBLEM

Step one: Identifying and clarifying questions, issues, problems

1 Identifying *central questions and issues*
2 Identifying *value questions*
3 Detecting *ambiguity* and *vagueness* of statement
4 Restating questions, problems, issues, in clear, precise, unambiguous terms
5 Identifying the elements of a question, problem, issue, in need of further *clarification*
6 Identifying the '*valuing*' elements in a question, problem, issue.
7 Identifying areas of *conflict* (especially conflicting values) in a question problem, issue.
8 Identifying areas in need of *investigation*
9 Distinguishing between *direct questions* and *hypothetical questions*
10 Formulating *hypotheses*
11 Identifying appropriate procedures for *testing hypotheses*
12
13

Step two: Gathering and organising data

1 Locating *information*
2 Locating *sources* of information
3 Using *data-gathering techniques* used by social scientists (e.g. sampling, surveys, questionnaires, interviews, content analysis)
4 *Selecting* appropriate data
5 *Classifying* data
6 *Summarising* data
7 *Recording and presenting* data
8 Selecting appropriate *techniques for treatment* of data
9 *Translating* data from one form to another
10
11

PROCESSING THE DATA

Step three: Interpreting data

1 Understanding *form* in which data is presented
2 *Retrieving basic information* from single data source
3 *Retrieving detailed/less obvious information* from single data source
4 *Retrieving complex information* from single data source
5 *Retrieving information* requiring use of *more than one data source*
6 *Comparing data* from different sources
7 Distinguishing between *fact and opinion/speculation*

8 Distinguishing between *specific facts and general facts* (empirical generalisations)
9 Distinguishing between *factual statements and conditional or hypothetical statements*
10 Distinguishing between *factual statements and value judgements*
11 Distinguishing between *factual statements and normative statements*
12
13

Step four: Analysing data

1 Recognising underlying *assumptions*
2 Following a *line of argument* (especially where this is from an unfamiliar/unconventional point of view)
3 Determining the *point of view* of author
4 Detecting *logical flaws* in an argument
5 Detecting unwarranted assertions, inferences, conclusions, etc.
6 Detecting *relationships*, e.g. causal, chronological, concurrent, etc.
7 Making warranted *inferences/extrapolations* from data
8 Making warranted *interpolations* where there are gaps in data
9 Drawing warranted *conclusions* from data
10 Making warranted *predictions* of trends, consequences, etc., from data
11 Discerning *factors which may affect the accuracy* of predictions
12 Formulating *hypotheses* to account for effects observed in the data
13
14

Step five: Evaluating evidence

1 Recognising *stereotypes and clichés*
2 Detecting *emotive* elements in presentation
3 Detecting *bias and prejudice* in presentation
4 Detecting *motive/purpose* in presentation
5 Detecting *persuasive techniques* used in propaganda, advertising, etc.
6 Distinguishing between *verifiable and unverifiable* data
7 Distinguishing between *relevant and irrelevant* information
8 Distinguishing between *essential and incidental* information
9 Assessing the *adequacy/inadequacy* of data
10 Assessing the *appropriateness/inappropriateness* of data
11 Determining the *consistency/inconsistency* of evidence
12 Determining whether *facts support* a generalisation/conclusion/inference

13 Assessing the *reliability* of data/sources

14 Recognising *limitations/qualifications* in the data

15 Distinguishing between *anecdotal evidence* and objective data

16

17

REACHING AND APPLYING GENERALISATIONS

Step six: Generalising

1 Detecting *common elements* in data

2 Detecting *relationships* in data which could lead to valid generalisations

3 Detecting *limitations/deficiencies/gaps* in data which could render generalisations invalid

4 *Modifying or rejecting hypotheses* on the basis of evidence

5 Formulating *valid generalisations*

6 Recognising *limitations/probability* factors in generalisations involving human social behaviour in a geographical context.

7 Recognising the *tentative nature of generalisations* involving human social behaviour in a geographical context.

8 Discerning *factors (e.g. change) which may affect the validity* of generalisations

9

10

Step seven: Applying generalisations

1 Suggesting *tentative solutions* to making tentative decisions in relation to questions/issues/problems.

2 Posing *alternative solutions* or decisions to questions/issues/problems.

3 Examining *relative merits* of alternative solutions or decisions to problems

4 Proposing suitable *courses of action* in relation to social problems in a geographical context.

5 Proposing appropriate *techniques* for reaching generalisations and finding solutions to questions/issues/problems of a geographical nature

6 Predicting *probable consequences* of a course of action/inaction

7 Identifying areas in need of further evidence or *investigation*

8

9

Step eight: Making value judgements

1 Formulating *reasoned value judgements*

2 *Defending* a value position

3 Examining the *implications* of alternative value positions

4 Suggesting resolutions of *value conflicts*

5

6

SOURCE: Piper, K. (1976) *Evaluation in the Social Sciences*, National Committee on Social Science Teaching, Australian Government Publishing Service.

Sometimes only parts of the sequence will be in a learning activity and sometimes all steps will be gone through. The steps in the process should not be seen as a definition of a learning activity but rather as a guide to what skills may be required in the course of a substantial period of study. One learning activity may involve data processing and interpretation, while another may require students to generalise from the data and identify solutions to the questions or issues and make value judgements in relation to alternative solutions. In the field study reported by Clegg, students worked through the complete sequence. To complete this account of Piper's enquiry skills, a general description of the seven broad areas identified in Figure 5.8 is given along with examples applicable to geography lessons.

Piper's sub-categories

Identifying and clarifying questions, issues and problems involves defining questions/issues/problems and formulating further questions and sub-sets of questions and/or hypotheses to be investigated. The initial activity is identifying the topic for study (e.g. On what grounds could a new town be justified on the Canterbury Plains of New Zealand?).

This could be achieved by (i) identifying and clarifying the elements of the question (e.g. What is a new town? Where did the concept originate and for what reasons? What has been the experience of new towns—especially as nodes of growth?), (ii) investigating the central issue (e.g. Why is such a proposal being considered in New Zealand? Why were the Canterbury Plains selected as the site?), (iii) identifying value questions (e.g. Should the government intervene and plan for the redistribution of people and industry?).

The next step, *gathering and organising* data is the process of obtaining the information on which to base the investigation, and putting it into a form suitable for further treatment. To investigate 'On what grounds could a new town be justified?' requires the collection of newspaper reports and other reference materials and if feasible, opinions collected by interview and questionnaire. The material would then be classified and summarised verbally, statistically or graphically.

The first stage in the next part of the enquiry, *processing the data*, is selecting from the data information relevant and necessary to the enquiry and sequencing it for subsequent analysis. This may involve (i) comparing data from different sources (government and private data and reports) and (ii) distinguishing fact and opinion (e.g. personal observations, governmental agency reports, party political statements).

In *analysing data*, relationships which will lead to valid conclusions are sought. This process may include (i) recognising underlying assumptions (e.g. about the social costs and benefits), or (ii) making warranted predictions of trends from data (e.g. to what extent is industry likely to increase after initial establishment?).

Evaluating the evidence is the process of assessing the quality of the data, and skills brought to bear could encompass (i) distinguishing between verifiable and unverifiable evidence (e.g. When, and how were certain statistics collected?) and (ii) determining whether facts support a generalisation (e.g. that displacement of industry to the South Island will curb the growth of Auckland City).

The final main category, *reaching and applying generalisations*, consists of generalising, decision making, problem solving and making value judgements. Generalising, as I have already shown at length, is the process of using evidence to make general statements about the issue being investigated; decision making and/or problem solving is the process of using data as evidence to reach or suggest solutions or resolutions; and in making value judgements evidence is used as the basis for formulating reasoned value positions on the question, problem, or issue investigated.

Use should be made of data so that generalised statements about new towns and the redistribution of industry can be made. The process of generalising involves the development of tentative answers to questions such as 'what reasons do different groups in society give for the building of a new town? Why?' 'What evidence is there to suggest that new towns stimulate a sustained growth of industry?' 'Who benefits from such a scheme?' An enquiry will often be taken to the stage of making a decision or judgement on the basis of the information an investigation throws up. It might be that alternative actions to bringing about a more even distribution of industry would be suggested or a detailed report drawn up on why a new town should or should not be built. Answers to questions will be put forward, general

statements put together. Answering questions, reaching a generalisation or making a decision may require a reasoned stance to be taken in relation to a heavily value laden issue as Step 8 of Figure 5.7 suggests. Here the value analysis procedures described in Chapter 4 may be useful.

A view of learning

The foregoing discussion of the knowledge and enquiry skills engaged in learning, fills out the view of learning implicit in my planning model and specifies the range of abilities which can be engaged in studying geography. The process of enquiry as an interaction among the four areas—knowledge, skills, attitudes/values and social skills—becomes a process of learning. Learning activities planned in the asking questions to reach generalisations model are likely to promote learning in that general sense as well as in more specific senses. I think such a view of the process of learning has much to offer the teacher organising and sequencing activities and should be placed prominently alongside the psychology of learning and what it tells us. In fact, I see this process model of learning as being of great value when set alongside my general pattern for planning. The work of psychologists like Piaget, Ausubel and so on apply at an even more specific level. For example, when we are grappling with the details of how to introduce topics, we will want to take (1) the concept of advanced organisers into account, or (2) decide whether to present the data in enactive, iconic or symbolic modes, or (3) conclude whether we need to use concrete examples and demonstrations as a way to link up with the likely level of children's mental development, or (4) recall Gagné's hierarchy and decide at what step in his hierarchy students are likely to be at. Do we need to present material to increase their ability to discriminate among concepts, or to expand and fill out their concepts and so on?

Paralleling planning and learning

I have chosen throughout to emphasise a procedure for planning which can be logically related to the process of learning. I see the emphasis on procedure to be more akin to process models of curriculum development (Stenhouse, 1975) than objectives models though I agree with Graves (1979) that one may best suit one situation and the other be most helpful in other circumstances. The circumstances seem to lie, as I hinted in the introduction, at the level of generality at which the planning is being undertaken. At general levels, general aims give a direction to planning school programmes and assist in the selection of content and skills while at the level of planning classroom learning activities, I think most often we are concerned with how and by what procedures—learning activities, teaching strategies—we will organise and present work to achieve a process of learning. I reiterate that this seems to me to be a reasonable resolution to the objectives/process debate. It is a resolution which seems to square with practice and represents an understandable change of concern as the level of curriculum planning changes.

Evaluation

Any models of curriculum planning whether dealing at the most general levels with the whole geography curriculum, or at more specific levels with instructional programmes for a term or a year, or with learning activities to be undertaken over a week or two, all include an evaluation phase.

This phase encompasses two ideas depending on the purpose of the evaluation. Student learning can be evaluated or assessed or the segment of work itself can be evaluated. The assessment of student learning, issues, styles and techniques is very fully covered by Marsden (1976) and hence those concerns will not be taken up here except to suggest that data response type questions are growing in popularity and are very appropriate for assessing students who have been exposed to geography in the way I have been suggesting. Figure 5.8 is an example of a data response question which may follow a unit of work on settlement introduced in the fashion outlined in Chapter 1.

The questions posed about the effect of particular events in the imaginary landscape sketched in Figure 5.9 give students a chance to use concepts and ideas previously developed in subsequent thinking. Daniel and Hopkinson (personal communication) suggest the following points in relation to questions (a) to (h) might be made:

(a) Settlement A is likely to expand and roads leading to it will probably be improved.
(b) A packhorse and later a wagon route is established from B to A, without any significant expansion of settlement at B.
(c) B will have an increased number of colliery workers and canal boaters. A jetty is likely to be established at E.
(d) With the coming of the railway, the coal trade will be diverted from the wagon route. Industrial growth possibly occurs at A, B, and the bridge. Port-related industries develop at E.
(e) In 1955 light industry is established at C, there is some decline of industrial functions at A, compensated by a growing service demand at D.
(f) Railway closure brings increased traffic congestion.
(g) C expands towards P.
(h) Cottages in B become vacation homes and new building is concentrated in A and D.

Readers may have additional suggestions.

Evaluation of curriculum projects, units and learning sequences has also now generated a considerable body of literature and it is distinguished by a number of schools of thought. I shall confine discussion to (1) some very basic points central to the literature, (2) a questionnaire I found useful and practical for evaluation and (3) a description of one model of curriculum evaluation which has some particularly pertinent and penetrating concepts in relation to evaluation.

Early literature

An elaboration of the conscious practice of evaluating teaching and the beginnings of a geographical literature on evaluation date back to the American High School Geography Project, a project which invested a considerable part of its resources in evaluation. The team's chief purpose in evaluating was to improve the geography courses then being designed by them. In effect, the evaluation was a formative one designed to improve the course while it was in the process of development rather than a summative evaluation designed to judge the effectiveness of the finished product. In day-to-day, year-to-year teaching, both formative and summative evaluation takes

Figure 5.9 The settlement pattern of villages in an imaginary landscape

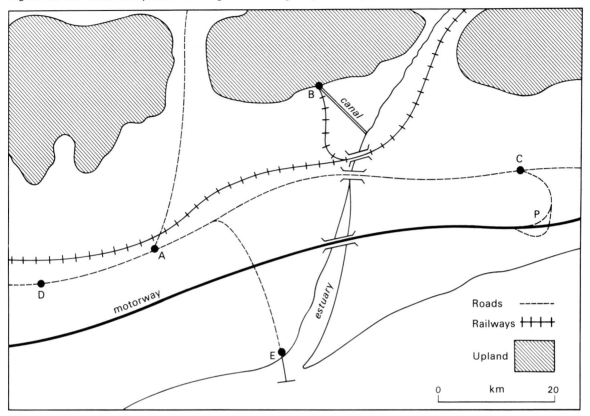

What effect will the following events have upon the pattern?
(a) the granting of a Charter in 1500 for a market to be held weekly at **A**.
(b) discovery of coal in 1700 at **B** in small quantities.
(c) building of a canal along the line indicated in 1785.
(d) coming of the railway along route indicated in 1856.
(e) establishment of a trading estate at **C** in 1955.
(f) closure of the railway in 1964.
(g) opening of motorway in 1970 along route indicated, with access at **P**.
(h) scheduling of upland area as greenbelt in 1975.
It is suggested that you redraw lines of communications and relative sizes of settlements
(**A, B, C, D**, and **E**) at each period.

SOURCE: Daniel, P. and Hopkinson, M. (1979) *The Geography of Settlement*, Oliver & Boyd.

place as modifications are made to teaching materials as a course progresses and before it is taught again. There was less emphasis finally in this Project on student performance and achievement, though some attention was given to this dimension because of the direct link back from student achievement to course improvement.

Many tasks, of which Figure 5.10 is an example, were initially devised to measure student learning. Students were given pre- and post-tests to determine whether learning activities had increased understanding. The scores for the questions in Figure 5.9 were:

	Pre-test	Post-test
1.	51%	72%
2.	45%	66%
3.	51%	69%
4.	44%	47%
5.	35%	28%

Figure 5.10 An HSGP student test item

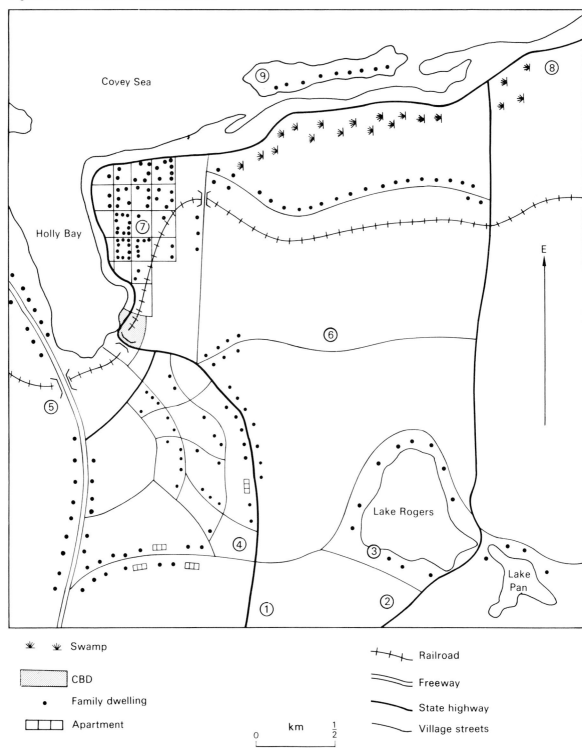

SOURCE: High School Geography Project.

Questions 1–3 refer to Figure 5.9

1 In which area of this community is there most likely to be low income housing?

A. 3
B. 4
C. 7
D. 9

2 Which of the following is the most probable location for a shopping centre designed to serve the mapped area for the next ten years?

A. 1
B. 3
C. 4
D. 7

3 In which of the following areas would you be most likely to find new factories and warehouses?

A. 3
B. 4
C. 5
D. 7

4 A surburban residential area developed during the 1970s would most likely be located near the crossing of

A. a railroad and a river.
B. a railroad and a major highway.
C. a river and a freeway.
D. two freeways.

5 Generally as the population of a city increases so does the

A. per cent of people involved in retail trade occupations.
B. frequency people go shopping.
C. percentage of income spent on food.
D. number of goods and services it offers.

Complexities in evaluation

The evaluators (Kurfman, 1970, 1972) came to acknowledge that the theoretically neat relationship among identification of concepts to be learned, materials and activities presented and test questions measuring learning could not so easily be established in practice. The complexity of ideas contained within the activities and materials made it difficult to unravel cause and consequence or identify what activity had contributed clarity or confusion, and therefore what precisely needed revision. Such tests came to be replaced by questionnaires.

This description somewhat brusquely highlights the difficulties inherent in using a scientific approach to try to evaluate such a complex process as teaching and learning and it is not surprising that new philosophies of evaluation have developed since the 1960s. One of the chief of these is illuminative evaluation.

Illuminative evaluation

Illuminative evaluation is concerned more with the overall process of innovation rather than just its outcomes. Illuminative evaluators are interested in noting and following up the effects of an innovation and making judgements on value and worthwhileness. The illuminative evaluator asks—'What happens when the innovation is introduced?' The experimental pre-test, treatment, post-test evaluator asks— 'Does the innovation perform as intended?' The teacher is probably intuitively aware of both questions and modifies exercises and courses in relation to his/her answers to both. If interest is heightened by new material or new methods, it is likely to be used again. If an arranged sequence of work achieves an intended progression of ideas it will be thought satisfactory. There may, of course, be some effects and outcomes which escape notice and outside evaluators may pick these up. An eclectic approach to evaluation is possible and the systematically gathered questionnaire data on which HSGP came to place most reliance is still a useful technique on which to base improvements to learning through geography even though it falls more into the scientific mode of evaluation rather than the illuminative school.

The use of questionnaires

Both student and teacher questionnaires were used to revise HSGP materials. The student opinions were found to be of great significance in comparing and deciding on the relative effectiveness of activities. Understandably, teacher opinions were of major importance in helping the unit developers make specific revisions. In the questionnaire data which yielded opinions on the worth, interest, enjoyment and effectiveness of activities, the most reliable factor in ranking activities was student interest. Student ranking of activities in terms of educational worth corresponded closely with their rankings in terms of interest. Teacher ranking of activities on any basis most often paralleled student interest rankings. Given the links between interest, motivation and learning, interest as a criterium in measuring course effectiveness seems legitimate. A brief summary of the exercise may serve to recommend it and the questionnaire method of obtaining information on learning activities and curriculum units. (Renner and Slater, 1974.)

Evaluating HSGP—the Japan unit

The Japan unit (HSGP, 1974) consists of four major activities including a role play. The first, an introductory unit, involves showing students a filmstrip of scenes of Japan and North America to enable students to note economic and cultural similarities and differences and to develop an awareness of Japan as a modern, industrial nation. The second activity consists of an essay on life in traditional Japan, centred around the customs and social habits of the samurai. The feudal structure of society and major characteristics militating against, or conversely facilitating change are conveyed.

In activity three, 'students contrast modern with traditional Japan and explore the problem of how a nation makes such a transition'. An essay on modern family life describes Japan today. The second section of this activity introduces the class to the concept of modernisation by using graphs and charts to illustrate some of the differences between developed and under-developed nations. The final section, which requires the largest amount of teaching time—up to six or seven periods—is an open-ended exercise. Students assume the role of experts in various aspects of economic development at a United Nations Conference set up to explore how under-developed countries might progress to development. The students must bring forward a set of recommendations. Committees study data to note the changes which occurred in agriculture, manufacturing, population and so on.

The evaluation sheet, Figure 5.11, was completed by twenty-four, 15-year-old New Zealand girls about four months after it had been used and at the end of the year's course. At the suggestion of one of them, a category 'much the same' was added to the otherwise HSGP style questionnaire. Experts hold that this is not a useful category. The majority of the class evaluated the unit positively.

The student interest in the unit was high. Part 2 of the completed questionnaire indicates that 87 per cent of the class found the activities to be interesting as a whole and over 70 per cent found the readings, graphs and Parts 2, 3, and 4 interesting. Fewer, 59.5 per cent, judged the graphs interesting and one pupil commented that the sheer number of graphs and other pictorial illustrations became 'monotonous'. Another comment stated, however, that 'the graphs helped [to learn] a lot'.

Showing the filmstrip was markedly the least successful activity with more than half the class not remembering it at all. The poor reproduction quality of the filmstrip may account for this. Of the majority of those who did remember it, none found it 'extremely interesting'. Comments in detail reveal: 'I thought . . . the readings were very lengthy considering the small amount of important facts they contained'; 'the most boring part of the topic was the reading as it did not allow for variation on the theme'; '. . . enjoyable to write essays [as the readings] meant there is a lot to "ramble" about'; and 'extremely interesting, not in the form of *straight* facts [facts included in stories of a family man's daily work etc.]'. More positive individual comments were made about the United Nations Conference as an opportunity 'to break away from the readings and graphs' and 'the United Nations Conference helped me to understand how a country changes and all the things it must do before it can even begin to start'.

Awareness of objectives

In assessing student awareness of objectives, the United Nations Conference ranks as the activity about which pupils were most 'clearly aware' of objectives (64 per cent). The reading on Yokichi, to bring out features of traditional Japan, was more clearly appreciated than the reading about the Japanese family man. The filmstrip again shows up as having the least student impact. The objectives of the work with graphs were known to 82 per cent of the students. In this section, the problem of assessing objectives from a skills or content point of view became most apparent. For example, a number of girls stated the objective as being to learn to read graphs and these kinds of responses were judged to represent a clear awareness of objectives. Over three-quarters of the class were aware of the objectives of the topic as a whole. It should be noted that three pupils stated that the main objective for studying Japan was to provide a topic for the School Certificate examination!

Estimation of student learning

Eighty-six per cent of the students estimated that they had learned quite a lot or a great deal from the unit and the reading, graph and conference activities were valuable learning experiences in the judgement of the majority of the class. The reading on traditional Japan is the leading section in the student estimate of their own learning (Part 4); the class was generally aware at the beginning of the exercise of Japan as an industrial nation. The filmstrip again shows up as a weak activity.

The kind of modification which would advisedly need to take place to materials and their presentation if taught a second time is reasonably clear. The benefits for student learning, planning activities and teacher performance seem to be considerable. As a final stage in planning learning activities, relatively formal methods of evaluation as opposed to informal methods are justified. If one were to ask the key questions characteristic of Simon and Wright's (1974) value strategy and sift through the evidence, then affirmative answers to 'Is it desirable?', and 'Is it feasible?' are most likely. The final step in planning activities has now been identified and discussed—namely assessment and evaluation.

Stake's model

There are, in addition, other perspectives from which to approach evaluation. Stake's (1967) model for curriculum evaluation can be applied conceptually to other planning levels and curriculum segments including relatively short lesson activities. One of Stake's major concerns is to try to achieve some description and judgement of what a teacher intends to do and what he or she does, i.e. a description of *intentions* and *outcomes* and the match between the two.

To accomplish such tasks a comprehensive plan is needed (Figure 5.12). Stake prescribes three areas in which data should be gathered which he labels antecedent, transaction and outcome data. An *antecedent* is any condition existing prior to teaching and learning which may relate to outcomes. Familiar antecedent conditions include student aptitude, previous experience, willingness and interest as well as resources available. *Transactions* include, for example, teacher to student, student to student, parent to student, parent to teacher and author to reader interactions and any others affecting the process of education. Class discussion, doing homework, using a video, marking work, completing an exercise, answering a question, all fall into the transaction category. *Outcomes* include the achievements, attitudes and aspirations of students and teachers, wear and tear on equipment, and impact on students, teachers and administrators, for example. Simply to view what is happening in a classroom as an interaction of antecedents and transactions contributing to outcomes is helpful and directly relatable to planning activities. To expand the concepts further and examine the relationships between intended and observed antecedents, transactions and outcomes, is a valuable evaluation strategy.

In planning the settlement siting activity for example, most teachers will first ask, 'What prior learning has been achieved in this field?' 'How easy or difficult is it going to be to settle students down?' 'How do I need to introduce the exercise?' 'What level of interest are students likely to display?' 'Do I need to try to motivate students?' 'How?' 'With visuals or a short initiating exercise as suggested in Chapter 1 on the location of settlements?' 'Do I work from an overhead transparency of the settlement site sketches or give everyone a copy?' 'Do students work individually or in groups?' 'Is it a written or oral exercise or if a combination how is this organised?' 'How do I judge outcomes?' 'Do I construct a test? Give homework? Sum up the generalisations myself? Record the decisions and generalisations from class discussion on the board as notes? Ask the students to make notes?' All these are quite small yet crucial decisions which need to be made in relation to antecedents, transactions and outcomes in planning a short learning activity. Wider questions obviously pertain to the concepts of antecedents, interactions and outcomes in a larger curriculum context but we can always ask, 'Have I planned with these three concepts in mind?' 'Did my plans work out?'

Contingency and congruence

Two additional concepts are important in filling out Stake's scheme and tying the first three concepts together—contingency and congruence. Contingency refers to unravelling the relationships among antecedents, transactions and outcomes and congruence to the general fit between intentions and outcomes.

Figure 5.11 Evaluating the Japan unit

1 *How did our study of Japan compare with our study of other topics in geography this year?*
☐ Much poorer   ☐ Somewhat poorer   ☐ Much the same   ☐ Somewhat better
☐ Much better.

2 *How interesting did you find our study of Japan?*

	Work as Whole	Reading Parts	Graphical Illustrations	Part One Filmstrip	Part Two Traditional Japan	Part Three Japanese Family Man	Part Four Modern- isation
Don't remember							
Dull							
Uninteresting							
Generally interesting							
Extremely interesting							

3 *What do you consider was the main objective (reason) for studying*
(a) *The topic as a whole?* To_____
(b) *The reading sections?* To_____
(c) *The graphs?* To_____
(d) *Part One?* To_____
(e) *Part Two?* To_____
(f) *Part Three?* To_____
(g) *Part Four?* To_____

4 *How much do you think you learned from*

	Nothing	Little	Quite a Lot	A Great Deal
(a) *The topic as a whole?*				
(b) *The readings?*				
(c) *Graphs?*				
(d) *Part One?*				
(e) *Part Two?*				
(f) *Part Three?*				
(g) *Part Four?*				

5 *Make any general comments, favourable or unfavourable, on any part of the work, if you wish.*
_____
_____
_____
_____
_____

SOURCE: High School Geography Project.

1 *Student Comparison of Japan Unit With Other Work (Number Responding 22)*

	Number	Percentage
Much poorer	1	4
Somewhat poorer	4	18
Much the same	5	23
Somewhat better	8	36
Much better	4	18

2 *Student Assessment of Interest*

	Work as a Whole	Reading Parts	Graphical Illustrations	Part One Film-Strip	Part Two Traditional Japan	Part Three Family Man	Part Four Modernization U.N. Conference
Number Responding	22	22	22	22	22	22	22
	%	%	%	%	%	%	%
Don't remember	4.5	9.0	0.0	55.0	13.5	13.5	0.0
Dull	4.5	4.5	9.0	4.5	4.5	9.0	13.5
Uninteresting	4.5	9.0	31.5	4.5	9.0	4.5	13.5
Generally interesting	77.0	64.0	46.0	36.0	46.0	46.0	50.0
Extremely interesting	10.0	13.5	13.5	0.0	27.0	27.0	23.0

3 *Student Awareness of Objectives*

	Topic as a Whole	Reading Sections	Graphs	Part One	Part Two	Part Three	Part Four
Number Responding	21	21	22	19	21	18	22
	%	%	%	%	%	%	%
Not Aware	18.0	32.0	0.0	32.0	9.0	18.0	18.0
Vaguely Aware	13.5	36.0	18.0	55.0	13.5	18.0	4.5
Generally Aware	13.5	4.5	32.0	9.0	68.0	41.0	13.5
Clearly Aware	55.0	27.0	50.0	4.5	9.0	23.0	64.0

4 *Student Estimate of Own Learning*

	Topic as a Whole	Reading Sections	Graphs	Part One	Part Two	Part Three	Part Four
Number Responding	22	22	22	22	22	22	22
	%	%	%	%	%	%	%
Nothing	0.0	0.0	4.5	27.5	9.0	4.5	0.0
Little	13.5	23.0	23.0	32.0	9.0	27.0	27.0
Quite a lot	89.0	77.0	59.0	36.0	68.0	64.0	41.0
A great deal	27.0	0.0	14.0	4.5	13.5	4.5	32.0

Figure 5.12 The elements in Stake's model of evaluation

Descriptive data

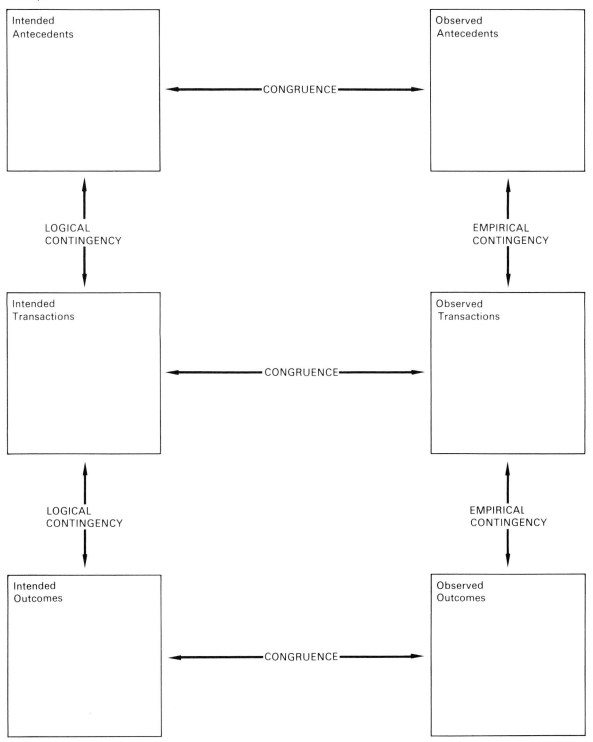

Source: Stake, R. (1967) 'The countenance of educational evaluation', in Taylor, P. and Cowley, D. (eds), *Readings in Curriculum Evaluation*, William Brown.

Congruence occurs if what was intended actually happens. This idea speaks loudly to teachers. If evaluation of teaching and learning activities takes place with congruence in mind, then it can act as a powerful idea for refining and modifying learning activities and sequences. Activities may be given another balance, some parts of the work may be dropped, others enlarged on, some may be presented differently and so on. To ask, 'Was there a congruence between what I intended, what I did and what I achieved?' gives a direction and focus to thoughts when evaluating and revising work.

What of contingency? Activity planning and curriculum revision is often based on a faith in certain contingencies. Presentation is arranged and resources selected to match intended learning. The probable connections or contingencies between them are often based on intuitive, subjective judgements which may be accurate. There is a case, however, for applying logic and asking, 'Is there a logical connection between an intended transaction and intended outcome?' 'Will showing a film on urban renewal (intended transaction) expose students to land use conflicts in the inner city (intended outcome)?' If such a logical connection exists, then a logical contingency exists between the two intentions and should be transformed in observation to an empirical contingency i.e. the logic is borne out by observation of what takes place in the classroom. If logical and empirical contingencies do not exist, then some revamping or reselection of resources and strategies is required.

The concepts Stake embeds within his curriculum evaluation would seem valuable in day-to-day and week-by-week planning and revision. They tie in closely with education and teaching as experienced and provide props for reflection in the planning and replanning of activities. Stake's model is intended for professional evaluators but there is no reason why teachers should not apply it as a guide in planning and evaluation. Coupled with questionnaire data from students, concepts like congruence and information like that on antecedents, for example, help to develop an awareness on which to base planning and undertake evaluation. I have therefore added a final undertaking to the scheme of planning activities—evaluation—and suggested ideas for and ways of making this practical.

Evaluating the planning model

I feel I should now give the reader an opportunity to evaluate the worth of the planning model which has been elaborated with examples, step-by-step throughout the book. The choice of criteria, based on Raths (1971), allows me to emphasise the kind of intentions, transactions and outcomes which are of value and importance, if all other things are equal and my assumptions have been at all reasonable. Generally speaking, all other things are not equal and perhaps some readers will feel I have suggested a way of planning which is normative rather than realistic. Such has not been my intention and yet I acknowledge the existence of circumstances outside my immediate teaching experience and knowledge. My specifications for planning may be evaluated against the following criteria. Does my planning procedure allow the criteria to be fulfilled?

1. All other things being equal, one activity is more worthwhile than another if it permits children to make informed choices in carrying out the activity and to reflect on the consequences of their choices.
2. All other things being equal, one activity is more worthwhile than another if it assigns to students active roles in the learning situation rather than passive ones.
3. All other things being equal, one activity is more worthwhile than another if it asks students to engage in enquiry into ideas, applications of intellectual processes, or current problems, either personal or social.
4. All other things being equal, one activity is more worthwhile than another if it involves children with realia (i.e. real objects, materials and artefacts).
5. All other things being equal, one activity is more worthwhile than another if completion of the activity may be accomplished successfully by children at several different levels of ability.
6. All other things being equal, one activity is more worthwhile than another if it asks students to examine *in a new setting* an idea, an application of an intellectual process, or a current problem which has been *previously studied*.
7. All other things being equal, one activity is more worthwhile than another if it requires students to examine topics or issues that citizens in our

society do not normally examine—and that are typically ignored by the major communication media in the nation.

8. All other things being equal, one activity is more worthwhile than another if it involves students and faculty members in 'risk' taking—not a risk to life or limb, but a risk of success or failure.

9. All other things being equal, one activity is more worthwhile than another if it requires students to rewrite, rehearse, and polish their initial efforts.

10. All other things being equal, one activity is more worthwhile than another if it involves students in the application and mastery of meaningful rules, standards, or disciplines.

11. All other things being equal, one activity is more worthwhile than another if it gives students a chance to share the planning, the carrying out of a plan, or the results of an activity with others.

12. All other things being equal, one activity is more worthwhile than another if it is relevant to the expressed purposes of the students.

Conclusion

To organise and assist the process of learning through geography is a complex undertaking for both teacher and student. I have attempted to give readers a general pattern of planning learning activities which identifies significant steps or elements.

I do not seek in any way to suggest that what I am saying is revolutionary, new or something not practised in part or in total by teachers. What I do seek is to give some guidelines which student teachers may find helpful and experienced teachers something they can think about critically and hopefully, creatively. If so, individual practice may move beyond its present articulation and so take my ideas on planning and how best to go about that essential task yet further and more effectively forward.

I should like to conclude not by summarising what has gone before but by addressing myself very briefly to a 'so what' question. I have assumed throughout this book that learning through geography has something to offer the education of people who will be making significant decisions in society in the twenty-first century.

The very title of my book begs the questions: 'Why learn through geography?' 'What has it got to offer individuals or society?'

I take one last learning activity which I would label as a geographical activity to suggest that the subject has something to contribute to the development of mind and to an awareness of our individual societal impacts and obligations.

Classroom activity
Planning for the future

Consider the future. What sort of a future do you want fifty years from now? You are an individual inhabiting a relatively small, cohesive, egalitarian society labelled New Zealand. You are going to make a series of resource demands and a series of population growth decisions. These will give you one of twelve possible futures which would result if a high proportion of the population made the same decisions as you. You will then identify the decisions that are essential to that future and the decisions that must be avoided if that future is to be a reality (Scott, 1978).

Your resource demands will be made from the following list:

	Cost	Resource score
Self-sufficient community	$40 000	
Large house in the country	$30 000	
New home in the suburbs	$25 000	
Small apartment near your work	$15 000	
Big station-wagon	$12 000	
Small family car	$8000	
Public transport	$3000	
Family set of bicycles	$2000	
Holiday home	$15 000	
Volunteer worker	$10 000	
Lifetime hobby or sport	$5000	
More education	$5000	
Speed boat	$10 000	
Household goods	$8000	
Heated swimming pool	$6000	
Motorbike	$4000	
Recycling centre	$10 000	
Energy conservation	$6000	
Low pollution suburb	$5000	
Solar heating	$4000	

Total cost

Total resource score

Each item which is portrayed on a set of cards (not shown here) has been costed and there is a limited sum to spend. Players are to put together a package to show the way they would like to spend their life's income (set at $70,000) and then refer costs to a resource score specified in Scott's package and based on calculated energy demands. The reasoning behind the costs set out are explained in Scott's Teacher's Guide.

Population growth decisions are then made by the students and relate to (1) preferred family size, (2) national immigration, and (3) national population policies. A second score, a population score, is obtained and added to the first resource demand score. The total score then links to a future. Figure 5.13 details two of twelve possibilities or scenarios based on Scott's preliminary and careful research. The examples detail futures contingent on a high and a low total score respectively.

Figure 5.13 Future number 4 and number 12

FUTURE NUMBER 4

This future has resulted from all members of our society co-operating to lower both our economic and population growth. We are *more* interested in preserving our environment and having plenty of leisure time. We are *less* interested in a high standard of living. As a result our standard of living has not grown since the 1970s. We are also very independent. We produce almost everything we need ourselves—including all our energy. The conservation of energy has become one of the most important features of our society.

New Zealand's population has grown slowly. There are now 3.5 million people in New Zealand—about the same number as there were in 1980. We have a balanced population structure with an even proportion of young people, workers and old people.

Because our population has grown slowly:
—There are plenty of good houses and schools.
—There is plenty of food (New Zealand is a food exporter).

Because we have lowered our economic growth:
—Our standard of living is the same as New Zealand in the 1970s.
—We have enough energy without nuclear power.
—We have plenty of leisure time.
—Our environment has been protected (including all our remaining native animals and plants).

SUMMARY
FUTURE NUMBER 4
Population size— moderate
Population structure— balanced
Economic growth— zero
Standard of living— like New Zealand now
Environmental quality— high

FUTURE NUMBER 12

New Zealand has serious social, environmental and financial problems, due to a dangerous combination of very high population growth and very high economic growth. There are now 9 million people in New Zealand (including over two million in Auckland). Our population is unbalanced because there are too many children and not enough schools. There are not enough houses. All the food produced by New Zealand farmers is eaten here; so there is none left to export. Instead, we have to import food. There is no spare timber either; so our only exports are manufactured products.

New Zealand is a highly industrialised society. Many factories work seven days a week and 24 hours a day, causing much pollution. Disposal of radioactive waste from our 70 nuclear power stations is a problem.

New Zealand has a heavy foreign debt due to high levels of importation and government spending on nuclear power plants. With high levels of taxation and unsatisfactory social conditions, there is much social unrest involving strikes and violent demonstrations. Our large police force has become a para-military organisation.

Our society is close to collapse.

SUMMARY
FUTURE NUMBER 12
Population size— very high
Population structure— unbalanced
Economic growth— rapid
Standard of living— high (like U.S.A. now)
Environmental quality— very low
Nuclear power plants— 70+

SOURCE: Scott, G. (1978) *The New Zealand Futures Game*, Joint Centre for Environmental Sciences and the Commission for the Future, New Zealand.

I shall not describe or outline the activity further. I have used it to imply that geography in the curriculum has to be planned and organised in such a way that activities like this one and some of its possible related outcomes are incorporated into curricula. For example, students hopefully:

1. Comprehend that there are a number of possible futures.
2. Comprehend that their 'life' decisions influence the future.
3. Comprehend the consequences of different value systems.
4. Analyse the relationship between population growth, economic growth and environmental quality, and
5. Evaluate the different courses of action open to them.

Geography may continue to deserve a role in educational systems if we plan our learning activities thoroughly and conscientiously and if we are able to respond to changing student needs and societal needs and demands.

I am in effect adding a dimension to Fairgrieve's (1926) much quoted doctrine:

> The function of geography is to train future citizens to imagine accurately the conditions of the great world stage and so help them think sanely about political and social problems in the world around.[3]

I would add 'and to appreciate their individual responsibilities and to know how to act within the political and social systems of which they are a part'. We want more than an ability to imagine, we now need a geography which contributes to a student's developing value system and which gives him or her the confidence and know-how to participate in the narrower and wider communities to which he or she belongs. Ever-cheaper international flights and television documentaries are making the old geography largely obsolete. Geography as a kind of vicarious experience has to move forward to geography as enhanced engagement and participation in environments, whether these are wilderness places, urban places, developed places or developing places in local, regional or global contexts. The engagement and participation may be with recreational or social justice intentions in mind. For example, it may be a question of exerting oneself over the protection of

open space or the design of new housing in one's own area.

Geography in education needs to develop intellectual and social skills which permit people who have experienced the process to feel that they *can* participate in societies where they consider they belong. Whether cognitive perspectives gained in the process of education in general lead them into such decisions is another matter. Training has to be a part of education even if not the only part and geography has to both *open* minds and *train* minds so as to contribute to an individual's ability to experience, enjoy, and participate humanely in pre-industrial, industrial or post-industrial societies (Verduin-Muller, 1978). Geography cannot alone accomplish such a broad aim but it has to be in tune with the needs of individuals and societies if it is to continue to deserve a place in education.

Further reading

My suggestions for further reading are confined to the three fields: teaching styles and philosophies, language and evaluation. The former two as yet do not have by any means a vast literature though both are likely to grow. The third is a substantial field from which I have tried to make a balanced selection.

Teaching styles and philosophies

Fien, J. (1980) 'Operationalizing the humanistic perspective in geographical education', *Geographical Education*, Vol. 3, No. 4, pp. 507–32.

Hickman, G., Reynolds, J., Tolley, H. (1973) *A new professionalism for a changing geography*, Schools Council of England and Wales.

Huckle, J. (1980) 'Classroom approaches: towards a critical summary', in Rawling, E. (ed.), *Geography into the 1980s*, Geographical Association.

Huckle, J. (ed.) (1982) *Geographical Education: reflection and action*, OUP.

Knos, D. (1977) 'Problems of Education in Geography', *Journal of Geography in Higher Education*, Vol. 1, No. 1, pp. 13–19.

Knos, D. (1977) 'On Learning', in Manson, G. A. and Ridd, M. K. (eds.), *New Perspectives on Geographic Education*, Kendall/Hunt.

Marsh, C. J. (1978) 'Using inquiry approaches in teaching geography', *Journal of Geography*, January, pp. 29–35.

Rawling, E. (ed.) (1980) *Geography into the 1980s*, Geographical Association.

Richardson, R. (1982) 'Daring to be a teacher', in Huckle, J. (ed.), *Geographical Education: reflection and action*, OUP.

Romey, B., and Elberty, B. (1980) 'A person-centred approach to geography', *Journal of Geography in Higher Education*, Vol. 4, No. 1, pp. 61–71.

Language

D.E.S. (1979) *Aspects of Secondary Education*, Chapter 6.

Martin, N., Medway, P. and Smith, H. (1973) *From information to understanding*, Schools Council Writing Across the Curriculum Project.

Slater, F. A. (1979) 'The role of language in the geography lesson', *New Zealand Journal of Geography*, No. 67, pp. 18–19.

—and Spicer, B. J. (1980) 'Language and learning in a geographical context', *Geographical Education*, Vol. 3, No. 4.

Williams, M., (ed.) (1981) *Language teaching and learning in geography*, Ward Lock.

Evaluation

Biddle, D. S. (1971) 'Reliability in examining in geography', *Geographical Education*, Vol. 1, No. 3, pp. 286–95.

Briggs, K., Riley, D. and Tolley, H. (1979) *Data Response Exercises in Physical and Human Geography*, Oxford University Press.

Cooper, K. (1976) *Evaluation, assessment and record keeping in history, geography and social science*, Collins ESL.

Graves, N. J. (1982) 'The Evaluation of Geographical Education', in Graves, N. J. (ed.), *UNESCO handbook for the teaching of geography*, Longman.

Hall, D. (1976) *Geography and the Geography teacher*, Allen and Unwin, Chapters 2 and 5.

Helburn, N. (1979) 'An American's perspective of British Geography', *Geography*, Vol. 65, part 4, pp. 327–33.

Jones, S., and Reynolds, J. (1973) 'The development of a new "O" level syllabus', *Geography*, Vol. 53, No. 3, pp. 263, 268.

Kouimanos, J. (1980) 'Some considerations in the evaluation of student achievement', *Geography Bulletin*, Journal of the Geography Teacher of New South Wales, Vol. 12, No. 2, pp. 44–52.

Kurfman, D. (1970) 'Evaluating geographic learning', in Bacon, P. (ed.), *Focus on Geography*, National Council for the Social Studies.

Marsden, W. (1976) *Evaluating the Geography Curriculum*, Oliver and Boyd.

Piper, K. (1976) *Evaluation in the Social Sciences*, National Committee in Social Science Teaching, Australian Government Printing Service.

Reynolds, J. B. and Stevens, G. (1974) 'Geography 14–18, "O" level paper', *Bulletin for Environmental Education*, October, pp. 5–11.

Salmon, R. B. and Masterton, T. H. (1974) *The principles of objective testing in geography*, Heinemann.

Senathirajah, N. and Weiss, J. (1971) *Evaluation in Geography*, The Ontario Institute for Studies in Education.

Spicer, B. J. (1970) 'Some aspects of assessment in school geography', *Geographical Education*, Vol. 1, No. 2, pp. 155–68.

Spicer, B. J. (1976) *H.S.C. Geography evaluation materials*, Sorrett.

Tolley, H. and Reynolds, J. B. (1978) *Geography 14–18*, Schools Council/Macmillan, Chapters 7, 8 and 9.

Whitla, J. H. (1976) 'Classroom-centred Evaluation: a humanistic approach for the Social Studies', *Social Education*, Vol. 40, No. 7, pp. 568–73.

References

Chapter 1 Identifying Questions to Plan Learning Activities

1 Collingwood, R. G. (1939) *An Autobiography*, OUP.
2 ibid.
3 Secondary Geography Education Project (1977)*SGEP-PAK*, 115/25, Geography Teachers Association of Victoria.
4 Farbstein, J. and Kantrowitz, M. (1978) *People in Places*, Prentice-Hall.
5 Simonds, J. O. (1961) *Landscape Architecture: The Shaping of Man's Natural Environment*, Dodge.

Chapter 2 Planning Learning Activities to Reach Generalisations and Decisions

1 Fairbrother, N. (1970) *New Lives, New Landscapes*, The Architectural Press.

Chapter 3 Reaching Generalisations and Decisions through Processing and Interpreting Data

1 Shadbolt, M. (1971) *Strangers and Journeys*, Hodder and Stoughton.

Chapter 4 Interpreting and Analysing Attitudes and Values

1 Blyth, W. A. L. *et al* (1976) *Curriculum Planning in History, Geography and Social Science*, Schools Council/Collins.
2 Allen, R. (n.d.) *But the earth abideth forever*, National Council for Geographic Education, Instructional Activity Series 1A/5–16.
3 Superka, D. P. *et al* (1976) *Values Education Sourcebook*, Social Science Consortium ERIC Clearing House for Social Studies/Social Science Education.

Chapter 5 Learning through Geography

1 Clegg, A. A. (1969) 'Geography-ing or doing Geography', *Journal of Geography*, Vol. 68, No. 3, pp. 274–80.
2 ibid.
3 Fairgrieve, J. (1926) *Geography in School*, University of London Press.

Bibliography

Alexander, J. W. (1963) *Economic Geography*, Prentice-Hall.

Allen, R. (n.d.) *But the earth abideth forever*, National Council for Geographic Education, Instructional Activity Series 1A/5–16.

American Geological Institute (1970) *Essences 1*, Addison-Wesley.

Association of American Geographers and American Sociological Association (1974) *Experience in Inquiry*, Allyn and Bacon.

Ausubel, D. P. (1960) 'The use of advance organisers in the learning and retention of meaningful verbal material', *Journal of Educational Psychology*, Vol. 51, pp. 267–72.

Barnes, D. (1976) *From Communication to Curriculum*, Penguin Education.

Barnes, D. and Todd, F. (1977) *Communication and Learning in Small Groups*, Routledge and Kegan Paul.

Bartlett, L. (1982) 'Systems in Teaching Geography', *Teaching Geography*, Vol. 8, No. 2.

Beaver, S. H., Goodway, K., Rodgers, H. B. (1980) 'Cannock Chase—The Midland Park', *Geographical Magazine*, Vol. 53, No. 1, pp. 21–35.

Biddle, D. S. (1976a) 'Paradigms in Geography: some implications for curriculum development', *Geographical Education*, Vol. 2, No. 4, pp. 403–19.

—(1976b) *Translating Curriculum Theory into practice in geographical education: a systems approach*, Geographical Education Monograph Series, No. 1, Geography Teachers Association of Victoria.

Blachford, K. (1972) 'Values and Geographical Education', *Geographical Education*, Vol. 1, No. 4, pp. 319–30.

Blyth, W. A. L., Cooper, K., Derricott, R., Elliot, G., Sumner, H., Waplington, A. (1976) *Curriculum Planning in History, Geography and Social Science*, Schools Council/Collins .

Brent, A. (1978) *Philosophical foundations for the curriculum*, Allen and Unwin.

Briscall, J. R. (1980) *From Space to Place in School Geography Teaching*, Unpublished MA dissertation, University of London Institute of Education.

Brown, L. A. and Moore, E. (1968) *The Intra-Urban Migration Process: An Actor Oriented Model*, Department of Geography, Ohio State University, mimeo.

Bruner, J. S. (1960) *The Process of Education*, Vintage Books.

—(1966) *Toward a Theory of Instruction*, Harvard University Press.

Christian Aid, (n.d.) *Priorities in Development*, mimeographed sheets.

Claval, P. (1978) 'The aims of the teaching of geography in the second stage of French secondary education', in Graves (ed.), *Geographical Education: Curriculum Problems in Certain European Countries, with special references to the 16–19 age group*, IGU, Commission on Geography in Education.

Clegg, A. A. (1969) 'Geography-ing or doing Geography', *Journal of Geography*, Vol. 68, No. 5, pp. 274–80.

Collingwood, R. G. (1939) *An Autobiography*, Oxford University Press.

Cowie, P. (1974) *Value Teaching and Geographical Education*, Unpublished MA dissertation, University of London Institute of Education.

—(1978) 'Geography: a value laden subject in education', *Geographical Education*, Vol. 3, No. 2, pp. 133–46.

Daniel, P. and Hopkinson, M. (1979) *The Geography of Settlement*, Oliver and Boyd.

Davies, G. (1978) 'Testing central place principles in practice: exercises based on census of distribution data', *I.L.E.A. Geography Bulletin*, No. 3, pp. 14–21.

Davies, W. K. D. (1967) 'Centrality and the central place hierarchy', *Urban Studies*, No. 3, Vol. 63, pp. 61–79.

Day, D. and Millward, H. (1977) 'Analysis of village siting and rural isolation from topographic maps', *Geoscope*, Journal of the Provincial Association of Quebec Geography Teachers, Vol. 10, No. 2, pp. 63–77.

D.E.S. (1978) *The Teaching of Ideas in Geography*, H.M.S.O.

—(1979) *Aspects of Secondary Education*, H.M.S.O.

Eisner, E. W. (1979) *The Educational Imagination*, Macmillan.

Elliott, G. G. (1975) 'Evaluating classroom games and simulations', *Classroom Geographer*, October, pp. 3–5.

Engel, J. (1980) 'Perceptual Geography in the educational process', in Slater, F. A. and Spicer, B. J. (eds.), *Perception and Preference Studies at the International Level*, IGU, Commission on Geography in Education.

Ewing, M. (1976) *Learning about housing and related issues: with specific reference to Metropolitan Montreal*, Unpublished M.Ed. monograph, McGill University.

Fairbrother, N. (1970) *New Lives, New Landscapes*, The Architectural Press.

Fairgrieve, J. (1926) *Geography in School*, University of London Press.

Farbstein, J. and Kantrowitz, M. (1978) *People in Places*, Prentice-Hall.

Fenton, E. (1966) *Teaching the New Social Studies in Secondary Schools*, Holt, Rinehart and Winston.

Fien, J. (1979) 'Towards a humanistic perspective in geographical education', *Geographical Education*, Vol. 3, No. 3, pp. 407–21.

—(1980) 'Operationalizing the humanistic perspective in geographical education', *Geographical Education*, Vol. 3, No. 4, pp. 507–32.

—and Hull, C. (1980) 'How well do you know your city?', *Classroom Geographer*, October, pp. 3–10.

—and Slater, F. (1981) 'Exploring values and attitudes through group discussion and evaluation', *Classroom Geographer*, April, pp. 22–5.

Gagné, R. M. (1970) *The Conditions of Learning*, Holt, Rinehart and Winston.

Gilbert, R. J. (1979) 'Image, Experience, and Personal Geography: some implications for education',

Geographical Education, Vol. 3, No. 3, pp. 399–406.

Gilg, A. (1978) *Countryside Planning*, Methuen.

Gill, D. (1981), 'Geography in ILEA', *Issues in Race and Education*, No. 32, p. 3.

Gill, D. (1981) 'Geography for the Young School Leaver—a critique', *Bulletin for Environmental Education*, October, pp. 35–9.

Gillespie, J. A. (1972) 'Analyzing and evaluating classroom games', *Social Education*, Vol. 36, No. 1, January, pp. 33–42.

Goring, R. T. (1977) 'A British National Park Model', *Sagt*, Journal of the Scottish Association of Geography Teachers, No. 6, pp. 44–53.

Graves, N. J. (1971) 'Objectives in teaching particular subjects with special reference to the teaching of geography', *Bulletin of the University of London*, Institute of Education, N.S. No. 23.

— (1979) *Curriculum Planning in Geography*, Heinemann.

—(1975, 1980) *Geography in Education*, Heinemann.

—(1980) *Geographical Education in Secondary Schools*, Geographical Association.

Group for Environmental Education (1973) *The Process of Choice*, 4, MIT.

Gunn, A. M. (ed.) (1972) *High School Geography Project, Legacy for the Seventies*, Centre Educatif et Culturel, Montreal.

Hall, D. (1976) *Geography and the geography teacher*, Allen and Unwin.

Hamelin, L. (1972) 'Image mentale et connaissance réelle: l'exemple du Nord', in W. P. Adams and F. M. Helleiner (eds.), *International Geography*, Toronto University Press, pp. 1048–50.

Hanson, R. (1980) 'The changing landscape', *Bulletin for Environmental Education*, No. 106, pp. 10–14.

Harvey, D. (1973) *Social Justice in the city*, Johns Hopkins University Press.

Helburn, N. (1968) 'The educational objectives of high school geography', *Journal of Geography*, Vol. 67, No. 5, pp. 274–81.

—(1977) 'The Wildness Continuum', *The Professional Geographer*, Vol. 29, No. 4, pp. 333–6.

—(1979) 'An American's Perspective of British Geography', *Geography*, Vol. 64, Part 4, pp. 327–33.

Hickman, G., Reynolds, J. and Tolley, H. (1973) *A new professionalism for a changing geography*, Schools Council of England and Wales.

High School Geography Project of the Association of American Geographers (1965, 1979) *Geography in an Urban Age*, Macmillan.

Hopkins, G. M. (1953) 'Pied Beauty', in *Gerard Manley Hopkins*, Penguin.

Huckle, J. (1976) 'A consideration of some curriculum problems involved in developing an environmental ethic within geographical education', Unpublished MA dissertation, University of London, Institute of Education.

—(1976) 'Values and attitudes—the geography teacher's new frontier', *Teachers Talking*, April, Thomas Nelson.

—(1980) 'Values and the teaching of geography—towards a curriculum rationale', *Geographical Education*, Vol. 3, No. 4, pp. 533-44.

—(1980) 'Classroom approaches: towards a critical summary', in Rawling, E. (ed.), *Geography into the 1980s*, Geographical Association.

—(1981) 'Geography and Values Education', in Walford, R. (ed.), *Signposts for geography teaching*, Longman.

Hurst, M. Eliot (1968) *A systems analytic approach to economic geography*, Commission on College Geography, General Series, No. 8, Association of American Geographers.

Knight, C. L., Buckland, J. F. and McPherson, F. (1973) *New Zealand Geography: A systems approach*, Whitcombe and Tombs.

Kohlberg, L. (1975) 'The cognitive-developmental approach to moral education', *Phi Delta Kappan*, Vol. 56, No. 10, pp. 670-7.

Kurfman, D. G. (1970) 'Evaluating Geographic Learning', in Bacon, P. (ed.), *Focus on Geography*, National Council for the Social Studies.

—(1972) 'Functions of evaluation', in Gunn, A. M. (ed.), *High School Geography Project, Legacy for the Seventies*, Centre Educatif et Culturel, Montreal.

Long, I. L. M. (1953) 'Children's reactions to geographical pictures', *Geography*, Vol. 38.

—(1961) 'Research in picture study', *Geography*, Vol. 46.

Long, M. and Roberson, B. (1966) *Teaching Geography*, Heinemann.

Lunnon, A. J. (1969) *The understanding of certain geographical concepts by primary school children*, Unpublished M.Ed. Thesis, University of Birmingham.

Lynch, K. (ed.) (1977) *Growing Up in Cities*, MIT and UNESCO.

Macaulay, J. U. (1979) 'The use of a skills bank and skills cards in promoting effective inquiry techniques in school geography', *Proceedings of the 10th New Zealand Geography Conference and 49th ANZAAS Congress*, pp. 361-3.

McConnell, W. F., Coopman, C. M. and Hoban, C. A. (ed.) (1979) *Studying the Local Environment*, Allen and Unwin.

Marsden, W. E. (1976) *Evaluating the Geography Curriculum*, Oliver and Boyd.

—(1976) 'Principles, concepts and exemplars, and the structuring of curriculum units in geography', *Geographical Education*, Vol. 2, No. 4, pp. 421-9.

Martin, N., Medway, P. and Smith, H. (1973) *From information to understanding*, Schools Council Writing across the Curriculum Project.

—and D'Arcy, P., Newton, B. and Parker, R. (1976) *Writing and Learning Across the Curriculum 11-16*, Ward Lock.

Martorella, P. H. (1977) 'Teaching Geography through value strategies', in Manson, G. A. and Ridd, M. K. (eds.) *New Perspectives on Geographic Education*, Kendall/Hunt, pp. 130-61.

Maslow, A. H. (1943) 'A theory of human motivation', *Psychological Review*, Vol. 50, pp. 370-96.

Metcalf, L. E. (ed.) (1971) *Values Education*, National Council for the Social Studies.

National Geography Curriculum Committee (1978) *Draft National Guidelines*, Department of Education, New Zealand.

—(1980) *G6, Skills in Geography*, Department of Education, New Zealand.

Peel, E. A. (1971) *The nature of adolescent judgement*, Staples.

Peters, G. and Larkin, R. P. (1977) 'Industrial Landscapes: perception and classification as learning activities', *Journal of Geography*, Vol. 76, No. 5, pp. 183-5.

Piper, K. (1976) *Evaluation in the Social Sciences*, National Committee on Social Science Teaching, Australian Government Publishing Service.

Pring, R. (1973) 'Objectives and innovations: the irrelevance of theory', *London Educational Review*, Vol. 2, No. 3, pp. 46-54.

Raths, J. D. (1971) 'Teaching without specific objectives', *Educational Leadership*, April, pp. 714–20.

Raths, L., Harmin, M. and Simon, S. (1966) *Values and Teaching*, Merrill.

Rawling, E. (1976) *Motorway*, Geographical Association.

—(ed.) (1980) *Geography into the 1980s*, Geographical Association.

—and Rawling, J. (1979) 'A planner in the classroom', *The Planner*, Journal of the Royal Town Planning Institute, Vol. 65, No. 4, pp. 103–5.

Relph, E. (1976) *Place and Placelessness*, Pion.

Renner, J. and Slater, F. A. (1974) 'Geography in an urban age: trials of high school geography project materials in New Zealand schools', *Geographical Education*, Vol. 2, No. 2, pp. 195–205.

Reynolds, J. B. and Stevens, G. (1974) 'Geography 14-18, "O" level paper', *Bulletin for Environmental Education*, October, pp. 5–11.

Rhys, W. (1972) 'The development of logical thinking', in Graves, N. J. (ed.) *New Movements in the study and teaching of geography*, Temple Smith.

Rokeach, M. (1973) *The nature of human values*, Free Press.

Scott, G. (1978) *The New Zealand Futures Game*, Joint Centre for Environmental Sciences and the Commission for the Future, New Zealand.

Secondary Geography Education Project (1977) *SGEP-PAK*, Geography Teachers Association of Victoria.

Shadbolt, M. (1971) *Strangers and Journeys*, Hodder and Stoughton.

Shepherd, I. D. F., Cooper, Z. and Walker, D. R. F. (1980) *Computer Assisted Learning in Geography*, Council for Educational Technology.

Simon, F. and Wright, I. (1974) 'Moral education: problem solving and survival', *Journal of Moral Education*, Vol. 3, No. 3, pp. 241–8.

Simon, S. B., Howe, L. W. and Kirschenbaum, H. (1972) *Values Clarification*, Hart-Davis.

Simonds, J. O. (1961) *Landscape Architecture: The Shaping of Man's Natural Environment*, Dodge.

Simson, R. (1976) 'How useful are mental maps? A brief look at perception studies', unpublished paper.

Slater, F. A. (1974) 'The use of models in classroom geography', *Classroom Geographer*, December, pp. 3–12.

—(1975) 'Questions and answers: implications for geographical education', *Geographical Education*, Vol. 2, No. 3, pp. 265–79.

—(1976) 'The Concept of Perception and the Geography Teacher', *McGill Journal of Education*, Vol. 11, No. 2, pp. 166–77. Reprinted in *Bulletin for Environmental Education*, No. 69, pp. 14–18, and *The Newsletter* of the Geography Teachers Association of Victoria, June, 1977.

—(1980) 'Language and learning in a geographical context', *Geographical Education*, Vol. 3, No. 4, pp. 477–87.

—(1982) 'Studying relationships and building models through the analysis of maps and photographic evidence', in Graves, N. J. (ed.) *Handbook for the teaching of Geography*, Longman.

—and Gallagher, T. (1979) 'Mixed ability classes and schooling as socialization', *Geography in Newham*, Vol. 1, No. 3, pp. 52–9.

—and Spicer, B. J. (1977) 'Objectives in sixth-form geography: towards a consensus', *Classroom Geographer*, December, pp. 3–10.

—and Spicer, B. J. (eds.) (1980) *Perception and Preference at the International Level*, IGU Commission on Geography in Education.

Spicer, B. J. (1974) *Planning for Western Port*, Cassell, Australia.

—and Achurch, M., Blachford, K., Stringer, W. (1977) *The Global System 4: Space for Living*, Jacaranda Wiley.

Stake, R. (1967) 'The Countenance of educational evaluation', in Taylor, P. and Cowley, D. (eds.) *Readings in Curriculum Evaluation*, Brown.

Stenhouse, L. (1975) *An Introduction to Curriculum Research and Development*, Heinemann.

Stevens, W. (1980) 'Using computers in geography teaching 2. A classroom project in the sixth form', *Classroom Geographer*, April 1980, pp. 3–10.

Storm, M. (1979) 'Some tentative thoughts on a taboo topic', *Geography Bulletin*, I.L.E.A., No. 6, pp. 5–7.

Superka, D. P., Ahrens, C., Hedstrom, J. E., Ford, L. J. and Johnson, P. L. (1976) *Values Education Sourcebook*, Social Science Consortium ERIC Clearing House for Social Studies/Social Science Education.

Sussex European Research Centre (1978) *Exploring Europe, Rural Depopulation*, Autumn.

Taba, H. (1962) *Curriculum development: theory and practice*, Harcourt, Brace and World.

Tilley, R. G. (1974) 'The application of semantic differentiation in the classroom', *Profile*, Vol. 7, No. 19, Part 1, pp. 29–43.

Tolley, H. and Reynolds, J. B. (1978) *Geography 14–18: A Handbook for School-based Curriculum Development*/Schools Council, Macmillan.

Lockard, J. D. *et al* (1980) *UNESCO Handbook for Science Teachers*, UNESCO and Heinemann.

Verduin-Muller, H. (1978) 'A conceptual model for curriculum planning in geography', in Graves, N. J. (ed.) *Geographical Education: Curriculum Problems in Certain European Countries with special reference to the 16–19 age group*, IGU Commission on Geography in Education.

Walford, R. (ed.) (1981) *Signposts for Geography Teaching*, Longman.

Walker, A. H. (1979) 'Monte Carlo Simulation as a teaching technique', *Geoscope*, Journal of the Provincial Association of Geography Teachers, Quebec, Vol. 13, No. 1, pp. 8–19, and also in *Geography Bulletin*, Journal of the Geography Teachers Association of New South Wales, Vol. 11, No. 4, pp. 140–7.

Ward, C. and Fyson, A. (1972) *Streetwork: The exploding school*, Routledge and Kegan Paul.

Watson, J. W. (1977) 'Values in the classroom', *Geography*, Vol. 62, Vol. 3, pp. 198–204.

Wellard, R., *et al* (1973) *Out of Site*, Rigby.

Welsh, A. M. (1978) *The contribution of geography to citizenship education*, unpublished thesis, University of London Institute of Education.

Wheeler, K. (1976) 'Experiencing Townscape', a special issue of the *Bulletin for Environmental Education*, No. 68, pp. 3–28.

Williams, J. T. (1977), 'Learning to write or writing to learn?', NFER.

Williams, M. (ed.), (1981) *Language teaching and learning in geography*, Ward Lock.

Wilson, J., Williams, N. and Sugarman, B. (1968) *An introduction to moral education*, Penguin.

Wise, J. H. (1977) 'Music and geographical education', *Journal*, Geography Teachers Association of Queensland, Vol. 12, March, pp. 1–22.

Wolforth, J. and Leigh, R. (1971, 1978) *Urban Prospects*, McClelland and Stewart.

Wolpert, J. (1966) 'Migration as an adjustment to environmental stress', *Journal of Social Issues*, Vol. 22, No. 4, pp. 92–102.

Index